VALUE-BASED CONSULTING

For Irena, Richard and Edward

Value-Based Consulting

Fiona Czerniawska

First published 2002 by
PALGRAVE
Houndmills, Basingstoke, Hampshire RG21 6XS and
175 Fifth Avenue, New York, N.Y. 10010
Companies and representatives throughout the world

PALGRAVE is the new global academic imprint of
St. Martin's Press LLC Scholarly and Reference Division and
Palgrave Publishers Ltd (formerly Macmillan Press Ltd).

ISBN 978-0-333-97113-0

This book is printed on paper suitable for recycling and made from fully managed and sustained forest sources. Logging, pulping and manufacturing processes are expected to conform to the environmental regulations of the country of origin.

A catalogue record for this book is available
from the British Library.

Library of Congress Cataloging-in-Publication Data
Czerniawska, Fiona.
    Value-based consulting / Fiona Czerniawska.
        p. cm.
    Includes bibliographical references and index.
    ISBN 978-0-333-97113-0 (cloth)
    1. Consuting firms. 2. Value analysis (Cost control) I. Title.
    HD69.C6 C924 2002
    001--dc21
                                                    2002020829

Formatted by
The Ascenders Partnership, Basingstoke

10 9 8 7 6 5 4 3 2 1
11 10 09 08 07 06 05 04 03 02

Contents

PART 4
Conclusions

Acknowledgements

As always, I'm enormously grateful to those people who were kind enough to find time to be interviewed for this book. I'm a firm believer that no one person or organisation has all the answers, so it's tremendously important to me to reflect others' perspectives as well as my own.

But I'd also like to express my appreciation to all the other people who helped make those interviews happen and in particular: Geoff Dodds at PricewaterhouseCoopers, Jennifer Abbott at Manning Selvage & Lee, Paul Clarke at Andersen, Wendy Miller and Christina Wallace at Bain & Company, Bill Murray at Differentis, Milsom-Mann at Mercer Management Consulting, Kate Cleevely at the Weber Group, Linda Tavano at Silverline Technologies, and Andrew Giangola at McKinsey & Company.

I'd also like to thank the Management Consultancies Association in the UK for their permission to re-use material from their journal, *Spectra*, in Chapters 10 and 14.

Finally, my thanks go to Stephen Rutt, Jacky Kippenberger and their colleagues at Palgrave for all their commitment and support, and to my husband, Stefan, for all of his.

1

Introduction: What Clients Want from Consultants and What Consultants Want from Clients

The Consulting Conundrum

It's not hard to find negative feedback about consultancy in the media:

> Their [the pure-play e-consultants] plight arguably has more to do with their own self-destructive behaviour. They weren't designing websites, they were building 'end-to-end' solutions to save a client from getting crushed ... 'You have more money than time,' they cooed, and billed out their legions of tattooed 20-something webheads at $400 an hour.[1]

> When consultants show up and root around inside your company, several things can happen. They may perform dazzlingly, saving you millions of dollars and pointing you in the direction of lucrative new markets. They could also run amuck, costing you tons of money and pushing your company to the brink.[2]

With comments like these, it comes as a jolt to remember, not only that most of the world's largest companies regularly use consulting firms, but that a large proportion of consulting work is repeat business. 'It's helpful to use consultants who already know their way around our organisation and have credibility with members of the board', said one client I talked to. 'If we're embarking on a high-risk venture', said another, 'then it makes sense to limit our exposure by working with a firm we already know.' There are clear advantages for the consulting firm as well: like every other sector, it's much cheaper to win work from an existing client than to acquire a new one; continuity translates into a deeper understanding of a client's business, and this, in turn, reduces the cost of sales.

But familiarity almost inevitably breeds contempt. It's a small step from the advantages just cited to a 'better the devil we know' attitude. Consultants may be an easy target in the press, but, talking to clients

1

certainly in the more mature consulting markets of North America and Western Europe, there's no doubt that the public criticism of consultants is mirrored by continuous, low-level dissatisfaction. It's not (yet) enough to stop industry in its tracks but, left unchecked, it could grow into a serious problem.

In *Blown to Bits*, Philip Evans and Thomas Wurster[3] argue that successful businesses of the future will share three characteristics. They will have 'richness', an in-depth understanding of their market and the core competencies they bring to it. The second characteristic is 'reach' – the extent and variety of the networks (supplier, competitor and customer) of which they are part. Finally, they will have 'affiliation' – an ability to overcome the traditional, cultural divide between suppliers and customers by repositioning themselves as their customers' 'champion'.

The best consulting firms excel in both of the first two areas. They have responded to clients' cumulative demands for world-class expertise by developing specialised skills in specific areas – masters of a small number of trades, rather than jacks of none. They have exploited the opportunities of the late 1990s to position themselves as key brokers in the scramble to create collaborative ventures and exchanges. But it's the third and last characteristic that has been a problem. It's not that there isn't a demand for 'affiliation' among clients. Payment by results and the involvement of consultants in implementing their recommendations are just two of the ways in which demand is already making itself felt, and the prevalence of phrases along the lines of 'working in partnership with our clients' in consulting brochures shows that firms are aware of it. But, historically, consulting firms have found the idea of affiliation an anathema to their own strategic goals: for them, playing the market has made better sense, allowing them to spread the risk of a downturn in a particular sector. Narrowing their focus to a small number of markets and services has also been the kind of decision which many consulting firms – built around the consensus and individual power-bases of a partnership – have found hard to take.

This is, I believe, the hub of the problem. If you put the client-consultant relationship under the microscope, it becomes clear that there are many areas where each side's objectives are different, and some where they are diametrically opposed.

Take one of the most obvious ones. Clients want access to highly-specialised knowledge: it's something they put again and again as their top priority when they come to hire consultants. Why? Because specialised knowledge is one of the most visible areas in which consultants can add value. A consultant who walks in with a wealth of insights culled from direct, practical experience, has a tremendous advantage over someone whose knowledge comes primarily from a training course, however good. But developing such in-depth expertise poses the consulting firm with two problems. First, specialised labour can be inflexible labour. If you've spent two years exposing someone to – let's say setting up call-centres for financial services clients – and the bottom drops out of that market, then the effort involved in retraining that person will be significantly greater than it would be for someone who's spent the same period moving from project to project and sector to sector. It's rather like teaching a foreign language: it takes longer to teach someone who only speaks one language (their native tongue), than someone who's learnt one (or even several) foreign languages before. The former not only lack the practice of learning, but they don't have the structure (for example, an understanding of grammar) that enables them to learn efficiently. Second, the payback on investment in specialist expertise is very uncertain. Highly qualified (and therefore valuable) individuals are prone to leave; downturns in the market can occur with little notice – as the last couple of years have demonstrated. From the perspective of the consulting firm, there's a trade-off to be made: you can either build a niche workforce that is capable of earning high margin fees in the short term, or develop more generalist people who may not be able to command the highest rates but are more easy to re-orientate as clients' needs change. In other words: you can keep your clients happy or you can keep your investors happy. Doing both is difficult.

Or is it?

About five years ago, I was involved as a consultant with a major European retailer which wanted to produce a quantum leap in terms of the level of customer service it provided – something it had tried before, but without success. Given a choice between serving a customer and checking inventory, store managers habitually chose the latter, because good service was seen as something that cost money and the store managers were rewarded according to the profitability they achieved. The key, we found, was to develop a blueprint of an organisation in which service and profits were not mutually exclusive. I still remember one of the stories we came across to illustrate this. A journalist decided to test out US retailer Nordstrom's reputation for excellent customer service, first by taking back a shirt he'd bought there the previous day (he got his money back, no questions asked), next by taking back a shirt he'd bought from a different chain (he again got his money back), and, finally, by taking back something that he'd not only not bought at Nordstrom but which Nordstrom doesn't sell (a tyre for his car – and he still got his money back). Stunned, he rang the store's manager to ask how he could do this without going bust. 'It's easy', said the store manager, 'when the store wasn't busy, we asked one of the junior sales staff to go back to the store you bought your second shirt from, and the car tyre store, and ask for your money back. We used up spare capacity to re-coup what we'd paid out but also delivered customer services which far exceeded your expectations.'

Now, I have no idea whether this story is true or whether it's one of those myths that sometime grow up around organisations, but I think it brilliantly illustrates the point I want to make. Consulting firms make trade-offs because they think they have no choice. Most large firms prefer to have more flexible people rather than compromise their long-term existence. Small firms tend to focus on specific areas, but accept that their shelf-life may be limited as a result. Some – often mid-sized firms – fall between the two stools. And it's these trade-offs that breed client dissatisfaction. The objectives or clients and consultants are mis-aligned: if one side wins, it's at the other's expense, and *vice versa*.

But that may not be the only way. The purpose of this book is to show how Nordstrom-like thinking could be applied to the consulting industry – how it might be possible to deliver the value that clients want without compromising the value that consulting firms wish to create for themselves and their investors. And it seems particularly pertinent to look at this subject now, in the aftermath of the first wave of e-business consultancy and as many of the new generation of consulting firms set up at the crest of that first wave are either failing or have been subsumed into more established firms. Many of the new firms grew on the back of their reputation for innovation: since their collapse, it's been tempting to see creativity – however attractive it was and is to clients – as inimical to profitability. At the same time, I don't wish to make the mistake of being too prescriptive. Every consulting firm – like every client – is different, balancing its own, unique combination of strategic choices. What follows, then, is more an attempt to start a discussion – about what the possibilities are, about how some firms could change the way they do some things.

The Client Perspective

So what do clients really value when it comes to consultancy?

Essentially, they want to hire the right firm to do the right project, delivered in the right way. Some consulting projects never recover from the initial mismatch. A firm that may be highly capable of doing a project in one field is hired to one in another where its expertise is much thinner on the ground – a round peg going into a square hole, either because the client hasn't been sufficiently clear about what is required or because the firm has managed to persuade the client that it is capable of delivering in this new field. Other projects may start well – with a real synergy between the client's and consultants' strengths – but come unstuck along the way, perhaps because the consulting firm doesn't generate the new ideas expected by the client or can't supply the level of in-depth expertise required. The internal management of the consulting firm may also have an indirect impact: its project management may be cumbersome, its culture bureaucratic. Other projects simply end badly: the client has got what it wants, but the manner of delivery has been unsatisfactory – the consulting team may have been disorganised, people may have had to be moved around at short notice, notional partners may not have been able to work together effectively in practice.

This book is divided into three main sections. The first looks at clients' first requirement – that they employ the right firm; the second, at what they want the consultants to do – the right thing; and the third, at the manner in which the consultants go about their work – the right way (Figure 1.1).

The right firm

Choosing the right firm will never be easy. There are too many variables, too many people involved for even the most sophisticated corporate purchasers of consultancy to be confident that they can find the best firm for the job all of the time. But, that being said, there are – clients certainly believe – things that could be done to increase the probability that they can identify the most appropriate firm.

A constant complaint is about the lack of precision with which consultants describe their services in their marketing literature and on their web-sites. As one client put it, 'we don't expect everyone to do everything. In fact, we think far more highly of a firm that is honest about what it can and can't do, than one that's pretending to be good

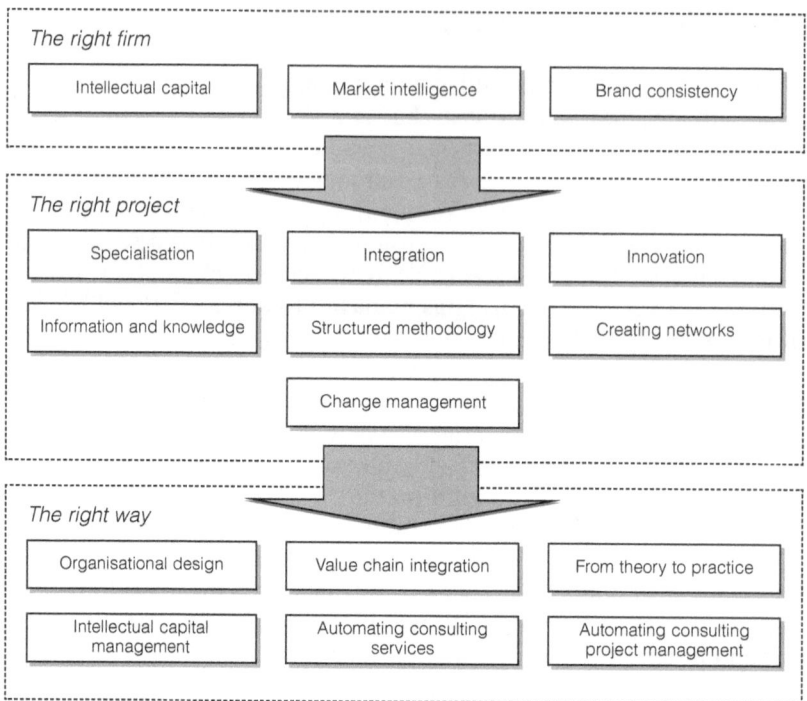

Figure 1.1 *What clients want*

at everything.' 'We want to understand how their skills dovetail with our own: all too often it's an all-or-nothing decision', commented another. 'Is it so very difficult for consulting firms to be clear about what they do and how they do it?' asked a third. It is hard – hard for any consulting company to be totally explicit about what a particular service or project may involve. Clients differ, and projects have to be tailored for their particular circumstances. Things change: with the best will in the world consulting projects don't always turn out the way consultants expect them too, let alone their clients. But underpinning these understandable obstacles lies the fact that it can be dangerous for consultancies to be seen to be too wedded to something, whether that's a specific service, its sector-based experience, or the way in which it works. Markets and clients move on, and consulting firms need to move with them. Hedging their bets by not being too precise in the way they present themselves makes good commercial sense, therefore.

Another source of frustration to clients is the frequency with which consultants come to talk to them without bothering to learn much about their business or the industry it's in. It's the consulting equivalent of junk mail: people coming in with a pre-set agenda of ideas and services they want to sell, irrespective of the needs of individual clients. Consultants would like to be better informed about what clients want, too, as an improving conversion rate of client meetings into billable projects won is one of the most effective ways of bringing down their cost of sales. But consultants also want to know about other projects in the offing. In other words, where individual clients want consultants who know a lot about their specific problem, a consulting firm wants to know about a lot of problems and is – implicitly at least – prepared to sacrifice how much it knows about any one problem in order to achieve this. To protect their margins, many firms, for example, have rules-of-thumb which demand that consultants limit the cost of sales to 5–10 per cent of the total project fees. It's yet another application of that almost ubiquitously useful 80:20 rule: consulting firms are prepared to invest 20 per cent of the effort to get 80 per cent of the understanding of a client's needs – but clients would rather it was 100 per cent.

Finally, there's the issue of whether the firm can live up to the marketing image it projects in reality. The 'experience' of consulting is as important for consultancies as the experience of a shop is for a retailer, or the experience of a theme park for its operator – perhaps more so, given the intangibility and perishability of the consulting 'product'. What clients want is a guarantee that what they see during the sales process is what they will get once the contract has been seen. Historically, these guarantees have largely focused on ensuring that the people who make the sales pitch actually do the work, but they now extend far beyond this. Brand has become a substitute for accreditation within the consulting industry. You know that a lawyer is qualified, because he or she has been through a demonstrable training process. No such widely-accepted qualification exists in consulting, and clients therefore use the fact that an individual works for a reputable firm as the only viable alternative – to the constant frustration of equally skilled, but independent consultants. A seamless transition from pre to post-sale is what consulting firms want as well. For them, brand also serves another function – it means controlling the outlying individuals of their organisation, the stars whose egos may make them difficult to work with, as well as the poorly-performing individuals who threaten to compromise the firm's reputation. The process is as important as the end result – the quality that clients seek has to be balanced with consistency.

The right thing

Leaving aside the rather negative reasons why clients hire consultants (rubber-stamping decisions already taken, for example) – the majority of which are, in my experience, apocryphal – we're left with seven positive reasons. However, even these, when we unpick them in order to understand more fully what clients want and why, include areas where the needs of clients and consultants do not match.

■ When you ask clients why they've brought in consultants, their answer almost always includes the need to access specialist skills which their organisation does not have. They may have other reasons as well, but this is always a central concern. Yet an area in which consultants add the most value is also – ironically – one of the areas of greatest potential conflict. From the client's point of view a specialist skill is a valuable skill: the more specialised it is, the more valuable it is. But from the consultant's perspective, specialisation can be synonymous with lack of flexibility. Even small, niche consulting firms find that a market that favours them one moment, may turn against them the next. Specialists are more likely to find their skills redundant than generalists – and they're harder to retrain.

■ But clients, like all consumers, want to have their cake and eat it. They may rate specialist skills very highly when it comes to buying consultancy, but they don't want these skills in isolation. Often such specialised skills are only truly valuable when they can be integrated with other, equally specialist skills. This isn't generalism by another name – clients want the translatability, the openness of general managers, but they want that paired with the in-depth knowledge of experts. If developing specialist skills without exposing themselves to short-term obsolescence is one challenge for consulting firms, then integrating these same specialist skills into coherent client projects – let alone coherent organisations – is a far greater hurdle. Overcoming it has been the undoing of many e-business consultancies and may yet undo some of their more established rivals.

■ Some clients genuinely want access to innovative thinking – they want to know that they're in the forefront of developing new ideas and challenging accepted ways of working. Consultants want this too: even consultants who've chosen to specialise in a very specific field will want to be seen to be leading edge within it. The conflict lies in who foots the bill. Paradoxically, clients want consulting firms to embrace innovation as part of their standard processes, and not bill them for it; in other words, clients want consultants to be efficient innovators. Consultants want clients who are willing to accept that innovation has a price tag.

- Creative thinking, however, is only part of the picture. In a world in which massive volumes of information can be transmitted in seconds at negligible cost, consultancies have a potentially important role to play in interpreting and aggregating this information, distilling the usable information from the white noise of data that surrounds their clients. But information may be a threat to the consulting industry, as much as an opportunity. It may be an asset that is prohibitively expensive for a consulting firm to develop – but, without it, consulting firms may find themselves 'disenfranchised' by companies – clients, even – that have access to superior data. And how can consulting firms protect their position in the intellectual value chain, when more and more of their 'knowledge' is being commoditised into 'information'?

- For some clients, too, creativity may be something they're willing to sacrifice in the name of speed and efficiency. For them, invention is potentially dangerous: they want a structured, tried-and-tested approach to delivery – which is why they've turned to an experienced consulting firm. Surely this has to be an area where clients' and consultants' aims have to be aligned? Surely everyone benefits from a disciplined approach? But discipline can become a victim of its own success: while some discipline may inject focus and momentum into a difficult project, too much may reduce the extent to which a project is implemented with an individual organisation's unique requirements in mind. The benefits go down, as the levels of efficiency go up.

- One of the legacies of the e-business boom has been the recognition, by client and consultant alike, that the latter's network of contacts may be a very valuable asset in its own right. As collaboration between diverse companies increasingly becomes the order of the day, consulting firms are finding themselves very well positioned to play the role of brokers, bringing potential partners together and facilitating the relationship between them. From the client's perspective, the benefits lie in the consultant's reach – the breadth of their networks. That's something that moves the consultant into a seat more traditionally occupied by an investment banker and, while the enhanced status that goes with this is clearly welcome to consultants, it also raises more difficult questions. How does a consultant charge for this kind of work? Will something that is so valuable to clients also become something that consultants – ironically – can't afford to provide?

■ Last, but not least, clients hire consultants to make things happen – things that their own organisations could not do by themselves at all, or within an acceptable time frame. Another of the lessons of e-business has been that 'proposition development' – the idea of treating a project as a discrete business venture, housed in a separate physical and cultural environment in which everyone (client staff and consultants) can focus exclusively on the matter in hand – has brought a momentum to delivery that many consulting projects in the past have lacked. This momentum has carried through into long-term management. The transition from strategy to execution – from theory to practice – has always been fraught with difficulties in the consulting arena: firms that excelled at the first lack the wish and/or skills to carry on to the second; those firms that came in second, inevitably seized the opportunity to pick holes in the recommendations of their predecessors. But the problem is not simply one of clients' making. Clients have followed where consulting firms have led, pigeonholing the latter into an exclusively strategic or operational role, and rarely letting firms migrate between the two. Part of the motivation behind this is undoubtedly unconscious, but an element springs from clients' desire to contain the influence of consulting firms. A firm that moves from strategy to implementation may well move into on-going management. Clients, therefore, want practical help, but only up to a certain point. How can consultants manage this? From their point of view, they may well want to be doing chargeable work on a long-term basis, but how do you prevent such relationships becoming unstuck because the client believes an imperceptible line has been crossed, that the continued involvement of the consultants represents an abdication of management responsibility?

The right way

And clients don't just want what they want – so to speak. They also want it to be done in the right way.

Increasingly – and this comes through again and again in the interviews included in this book – clients want consulting firms that can 'walk the talk', who reinforce the theory they're trying to sell with a concrete track record of successfully applying such theory to themselves. Why buy knowledge management advice from a firm that can't demonstrate that it can manage its own knowledge effectively? Why ask a consulting firm to redesign your organisation if it patently can't develop an appropriate structure for itself?

It is becoming harder for a consulting firm to keep its internal and external activities separate, to keep the private from the public. The pressures for this are partly driven by the financial markets – as more firms are floated, financial accountability is being matched by managerial transparency. But they're also the result of chronic client dissatisfaction which, justly or not, sees consultants as proffering medicine they're reluctant to swallow themselves.

Historically, there's always been a tremendous cultural divide, so far as consulting firms are concerned, between the consultants and non-consultants. The consultants have hotlines to clients that provide them with an unrivalled source of authority with which to justify their decisions (which is one of the reasons why managing partners often continue to work with clients alongside their internal, operational responsibilities). By contrast, the behind-the-scenes people, without the legitimating authority of being able to say 'this is what clients want', have been largely sidelined. But in this brave new world of do-as-you-would-have-your-clients-do consulting, this cultural abyss has to be bridged.

Thus, the intricacies of aligning the goals of clients and consultants extend beyond the visible persona of the firm – beyond the way in which it sells itself, beyond the reasons why clients hire it – to the way in which the firm is managed.

■ *Organisational design.* If e-consulting showed anything, it was the slowness with which many incumbent consulting firms reacted to a challenge that involved them bringing different parts of their organisations together. When you get senior consultants to talk about the role of these new entrants, they tend to deny that anything has changed as a result of their temporary incursions. But dig a little deeper and you find that they're in the process of trying to replicate the organisational design and culture of these firms.

- *Value chain integration.* Of course, it's not just the e-consultancies that have forced this change. The whole concept of the extended organisation, or seeing companies as portfolios of customer opportunities and resources that can be deployed on a project by project basis, is changing the way in which consulting firms look at themselves. The challenge here is flexibility without chaos: organisations capable of reinventing themselves can also destroy themselves in the process.

- *Managing the transition from theory to practice.* While new, more networked organisational models may bring the flexibility and responsiveness required to meet clients' needs in the future, they may also have disadvantages. Greater fragmentation may magnify existing discontinuities in the consulting process; more interfaces may result in higher costs. A key issue here is how business strategy is translated into technology: conventionally, consulting firms have been good at strategy or implementation, but not both.

- *Knowledge management.* And more complex organisational structures, less cohesively knitted together, raise inevitable questions about the future of knowledge management in consulting firms. How will the structured, 'manageable' systems of the present handle the less structured, tacit information that such organisations will require?

- *Automating consulting service.* Clients want efficient delivery of consulting services, and evidence of that efficiency. But how far are they prepared to go? Is online consulting a step too far, sacrificing too substantial a part of the client-consultant relationship for minimal benefits? And if they saw a significant price differential between online and offline services, would that change their attitude?

- *Automating delivery.* 'Self-service' consulting is only one side of the possible benefits that technology may bring consulting firms in the future. Often lagging behind their clients in their adoption of new systems and applications, will automation of delivery and project management provide the transparency and flexibility that clients will demand in the future?

The structure of this book follows these three areas – the right firm, the right thing, the right way. As well as my own research, it brings together the views of different consulting firms – large and small, established and new – in addressing the problems highlighted. In doing so, I've attempted to create a very tentative blueprint of how consultants and consulting firms could – and in some cases, are – managing to overcome the more insidious problems that have attacked the consulting industry, and reconciling goals where reconciliation has been assumed to be impossible – in order to create greater value for clients.

---

[1] Ralph King, 'The Talented Mr Greenberg: The Story of Scient and the E-Consulting Bubble', *Business 2.0*, May 2001.

[2] Ronald Lieber, 'Hire a Consultant – and Start Praying', *Fortune*, August 1997.

[3] Philip Evans and Thomas S. Wurster, *Blown to Bits: How the New Economics of Information Transforms Strategy* (Harvard Busines School Press, 2000).

Part I
The Right Firm

2

Intellectual Capital: Articulating Your Portfolio

I believe we're at the start of a period that will see the industrialisation of the consulting industry ... Consulting services will become more tangible, more 'productised' than they have been in the past. Part of this is a reaction to greater scrutiny by the market than has been the case historically in this sector: when you find yourselves being compared to a 'traditional' company with physical products, you have to be very clear about what your assets – intellectual as well as physical – are.

Alan Buckle, COO Europe, KPMG Consulting

'What frustrates me', said one client I interviewed recently, 'is how difficult it is find out what a particular consulting firm is good at. It used to be that everyone said they were good at everything: now firms recognise that such claims are actually counterproductive, and they talk about only working in areas where they're strong. But this switch in marketing hasn't made much of a difference in practice. Consultants are always optimistic about their ability to do things they haven't done before, and that really muddies the water from the client's point of view.'

> *What clients want:*
> To have a clear understanding of the areas in which a consulting firm excels so that it can use the firm in that area and only in that area

> *Value-based consulting:*
> Being precise about the client and consultant's intellectual capital in order to generate long-term client trust and to avoid falling into the trap of short-term management bandwagons

> *What consultants want:*
> To be able to diversify, i.e. manage their intellectual capital in such a way that the peaks and troughs of demand are minimised

17

Clients want to pigeonhole consulting firms. They want to be able to look at what they're trying to achieve, identify the areas where they don't have the resources, time or simple inclination to do it themselves and find consulting firms who can do the work for them. This desire increases with a client's confidence. In the earliest days of any new management idea, no one – client or consultant – knows what's involved, exactly what skills and tools it will require. When clients ask for consultancy it's therefore almost always non-specific, and they're happy to deal with a consulting team of bright people capable of working things out as they go along. But as clients' experience in an area grows, so does the precision with which they can stipulate their requirements. Markets for consulting services don't mature because the services themselves mature, but because clients do. Indeed, the size of any given consulting market isn't – as we often tend to think – simply determined by the number of clients buying that service, but also by the sophistication with which they can pinpoint exactly what they need – the more precise their requirements, the more likely it is that they will be able to use in-house resources and the less likely that they will need a consulting firm to cover lots of areas – so the smaller the market (Figure 2.1).

The speed with which clients understand what is involved in adopting a new business idea is, in turn, determined by the amount of information to which clients have access. Thus, in the last ten years, the Internet has produced a veritable explosion of data sources and the speed with which a client becomes sophisticated – and a consulting market matures – has increased significantly. Clients themselves are aware of this. A recent survey commissioned by the Management Consultancies Association in the UK showed that two thirds of the clients surveyed believed it was important to gather information about a consulting firm's entire portfolio – primarily via the Internet – before finalising any purchase decision. Not surprisingly – given the wealth of potential information – a third of those surveyed were also looking at how the process by which they purchased consultancy could be streamlined using an e-procurement system. This flies in the face of much prevailing thinking within the consulting industry itself, where the possibility that a client's decision-making process could be supplemented, if not in part superseded, by web-based technology (as other organisational purchase processes have been), has been downplayed almost universally.

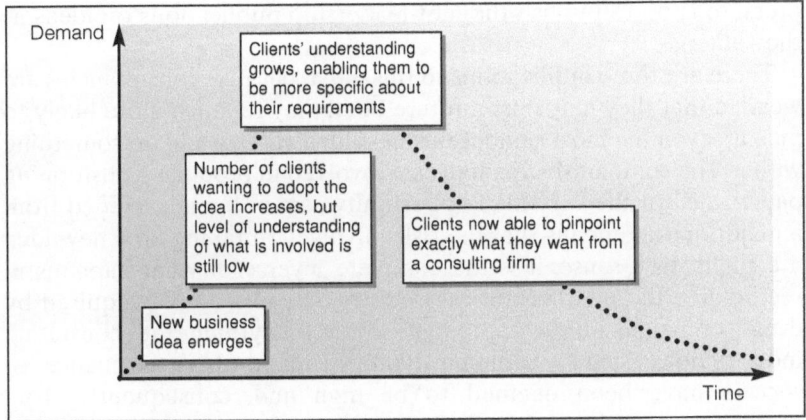

Demand

Clients' understanding grows, enabling them to be more specific about their requirements

Number of clients wanting to adopt the idea increases, but level of understanding of what is involved is still low

Clients now able to pinpoint exactly what they want from a consulting firm

New business idea emerges

Time

Figure 2.1 *Market size as a function of changing client demand*

More Precision, Less Speed

From the client's perspective, the first step in moving towards 'value-based' consultancy is for consulting firms to define their intellectual capital much more precisely and to demonstrate how that intellectual capital may (or may not) dovetail into the client's own requirements. In other words, clients want consulting firms to help speed up their learning process, enabling them to move more quickly from the early phase of a new idea (when you have to throw money at it) to the point where you know exactly what you want (and only need to buy that).

It doesn't take a degree from Harvard to see why consulting firms aren't exactly bending over to do this: faster maturing markets mean lower fees. There are other problems, too, from the firm's point of view. 'Clients have surprisingly long memories', one senior consultant told me. 'If you've been slow to get into a particular market, they won't let you forget it'. The sudden boom of the e-business consulting market in the late 1990s reinforced this imperative. Credibility in the consulting market has almost everything to do with having the right resources, and being able to respond quickly to an emerging market is still seen by clients to be a sign that a firm has at least some of these resources. The slow response of the incumbent firms to the boom in e-business demand was widely interpreted by clients as proof that these firms lacked the specialist skills now required. Stung by this experience, consulting firms now recognise that they have to trawl continually through the market in order to spot new markets: the pressure is on to publish thought-leadership publications on ideas as they emerge.

There are three implications to this. First, because consultancies are worried that they may miss a future boat, they're much more likely to pick up even the most slender business idea and try and do something with it. The comparatively small cost involved in producing positioning papers against the very high opportunity cost of being excluded from a major market means that the risks involved in taking up a new idea are slight. In a sense, it is the complete inverse of what happens in sectors like the pharmaceutical industry. The investment required by drug companies in new products dwarfs anything the consulting industry does. Such investment can only be made where the chances of success have been deemed to be high and, consequently, drug companies have highly developed methods for evaluating potential winners. Consulting firms, because the level of investment has historically been low, can afford to take more risks: the methods most firms have for selecting – and, indeed, de-selecting – new services for the future are far from comprehensive. The result is simple – far more new management ideas hit the market than new drugs.

The second implication makes matters even worse. Firms' positioning papers are written at an increasingly early stage in the evolution of the business idea itself. Rather than wait for the idea to take shape, consultants have to second-guess its future form and direction. Rather than commit to anything too specific, they hedge their bets and disguise their own uncertainty by assembling research, commissioning proprietary surveys and remaining suitably vague about what precisely is involved. There's rarely anything in these positioning papers that helps clients understand what they need, and what consultants can do. But the third and final implication is probably the most serious. Risk aversion at the front end of a market (meaning that consulting firms want to have coverage of more ideas at an ever earlier stage in their evolution) is matched by risk aversion at the back end. Why jettison an idea just because the market for it is declining, when there are still some fees to be earned? Nothing is lost by keeping an idea. The portfolios of consulting firms are essentially cumulative – they add, but they don't subtract. We tend to see the minimal cost of information (once created) as a major advantage over physical products. After all, it's this advantage that enables software companies to have apparently limitless scalability – the software (once developed) can be distributed to ten thousand people or ten million for just about the same amount of money. By contrast, a physical manufacturer – that bastion of the 'old' economy – has to buy raw materials, storage space, a distribution network in proportion to its sales. But perhaps the physical economy has one underrated advantage: finite capacity means that you don't carry all your products all of the time. Consulting firms have no such limitation: there's absolutely no economic incentive for them to ditch a service just because it's coming towards the end of its natural life: it doesn't cost anything to move or store it. Its only cost is in distribution (delivery) which can be matched against fees earned. Only when the fees clients are willing to pay have fallen below the costs of delivery does it make sense to discontinue the line.

Clients may want consulting firms to define the intellectual capital more precisely, but consulting firms have been doing just the opposite – and for sound economic reasons.

So, what's the solution? How can we – indeed, can we – reconcile the conflicting needs of each side?

A first step, I'd suggest, is to unpick some of the assumptions that underpin the situation as I've just described it:

■ Defining your intellectual capital more precisely only helps clients become more sophisticated more quickly, and that cuts into your sales and margins.

■ As a consulting firm, you can't afford to miss new management ideas as they emerge; if you do, your competitors will win.

■ The only way to make sure you leave all your options open in an emerging market is to be vague about what your proposition is.

■ There is no downside to keeping service offerings on the books, up to the point where they become manifestly unprofitable to deliver.

There's a simple equation underneath all of this: the more precise you are, the lower your fees will be (Figure 2.2).

Can this mindset be turned around, so that consulting firms can be seen to benefit from the precision their clients want? Can something – precision – that is valuable for clients also be of value to the consultants with whom they work? The most obvious – and least productive way – to say 'yes' is at the level of an individual firm. If your competitors are being vague about their offerings in a particular market, being precise is an important potential source of differentiation. Although the overall market size may decline – as your competitors follow suit – it's likely that you'll have won an

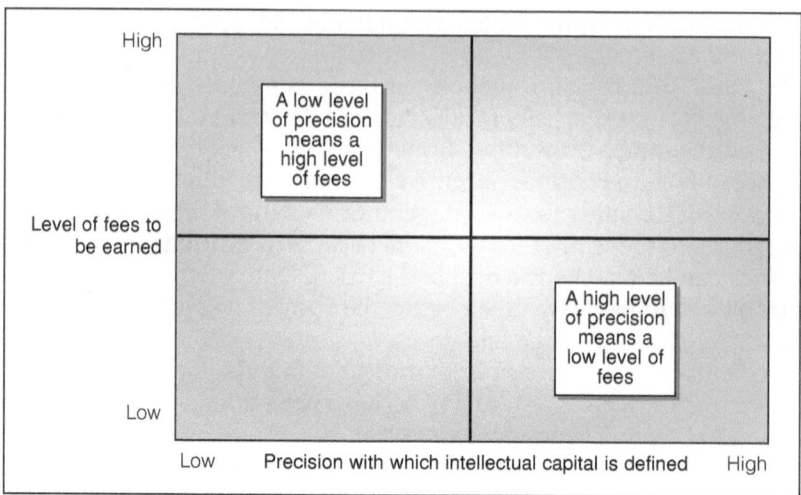

Figure 2.2 *A working assumption: precision and fees are mutually exclusive*

increased share of that business by being the first-to-precision rather than first-to-market. But, if everyone replicates your greater degree of preciseness quickly then all that happens is that you collectively cannibalise your own market: everyone earns less. To find the real benefits, you have to look more broadly at the overall relationship you have with your clients:

- Precision breeds trust. If a client can see that you've been very exact about what you can and cannot do in one area, they're more likely to believe what you say when it comes to what you can and cannot do in other areas. They're also more likely to listen to you when you talk about a new idea – providing, of course, you're being precise about it.

- Precision puts the client in control. Like all consumers, clients want to know what they're getting: being clear about what exactly you offer enables clients to decide what they need where – they don't have to rely on you making that judgement on their behalf (which is essentially what happens when clients don't really know what they need or what a consulting firm can provide). When you talk to them, almost all clients say that they'd prefer to put in more management effort on their side in order to assemble teams of genuine experts from different firms than cede this role to a firm that claims to offer everything.

■ Precision means that consulting firms have to rely less on management bandwagons. The pattern by which management ideas are disseminated and adopted increasingly resembles the stock market: sudden peaks of demand are followed by equally sudden falls. As consulting firms have become more astute about gathering market intelligence and identifying emerging ideas, they have become much more responsive to the behaviour of other firms – selling services others are selling. Such bandwagons create their own problems, stimulating demand among clients who think they should adopt the idea because their rivals have already done so, not because they would benefit from it. The more cohesive market activity is – the more that consulting firms sell the same 'big idea' at any one moment – the more likely it is that more clients will become more disaffected more quickly. Why? Because the number of clients who are following their competitors forms a greater proportion of the total number of clients. Management ideas, like rolling stones, gather moss. The bigger an idea gets, the more important it looks to potential clients. And, as the number of clients who adopt the idea grows – for the wrong reasons – the lower the average benefits will be and the greater the risk of a sudden drop in demand. Being clear about your exact service offering breaks this cycle: greater differentiation is injected into the market which, in turn, makes it more difficult for a single idea to gain the momentum required to metamorphose into a significant bandwagon. Breaking down potentially monolithic consulting markets may cut into fees in the short run, but it leads to more sustainable, although smaller, markets in the long run.

Precision, Not Prescription

So it is possible to align the economic aims of consulting firms with the needs of clients. But doing this in practice isn't exactly painting by numbers: how can you describe your services more exactly without compromising your flexibility to customise what you do for individual clients?

Until March 2001, Mark Curtis was the Executive Vice President for Global Learning and Skills at Razorfish. He's now in the process of launching a new company, Fijord.

❛ I think many of the specialist e-consultancies, founded since the mid-1990s, ran into problems because the definitions of their service offerings were too complicated. They became fixated on the idea of providing an end-to-end solution, doing everything from the upfront analysis of customer behaviour to the implementation of highly specialised technology at the back-end. It was both too grandiose a vision and one which most firms weren't equipped to deliver, especially in different parts of the world. It's a scale issue, as much as anything, in my view. Consulting firms that make it past the crucial hurdle of 40–50 people often start to believe they can do anything: when you're still only 20–30 people strong, you're far too focused on day-to-day survival and doing what you know best to have any broader ambitions.

And I don't think end-to-end consultancy is what clients want. The only person in an organisation capable of buying end-to-end solutions is the Chief Executive. Everyone else is managing parts of the organisation: only the Chief Executive is in a position to see everything. Given that the only consultancies that tend to have enduring relationships at this level – the top strategy consultancies – aren't trying to sell end-to-end services, perhaps Chief Executives don't want them either. For most other consulting firms, selling end-to-end solutions means that you have to sell to several people at once, each of whom is buying the component that relates most closely to their own work and objectives. The sales process gets longer and more expensive.

After leaving Razorfish, our first priority was to work out what of the last few years wasn't destroyed in the car wreck of 2000–01. There are some success stories: eBay; Amazon – probably; both Buzz and Ryanair say that the Internet is fundamental to their business models; corporations like GE are claiming significant savings by using web-enabled technology to streamline their business processes. But such stories are few and far between: it's become fashionable to see the Internet as just another channel. But its impact has been huge – it's just that it's also been in places where we don't tend to look for it. It's been in the social consequences of connectivity, rather than commerce. My thirteen-year-old daughter has a mobile phone she uses to send text messages to her network: some of the people she chats to she's met, others she just knows via her telephone. What kind of expectations about working and living will she have by the time she's twenty?

Take the workplace: changes have been brought about that were completely unthinkable ten years ago. One story I heard was about a company whose employees were regularly posting derogatory notes about it on an Internet site. The CEO, once he'd heard about it, asked the IT department to prevent people from accessing the site, at least while they were at work. Within hours everyone in the company had got to hear about this, creating a groundswell of negative feedback. None of these things could have happened when people weren't so connected: there wouldn't have been a web-site; people wouldn't have got a kick from posting comments on it; their colleagues wouldn't have started reading; the CEO wouldn't have been able to sever the connection; gossip about the CEO's response wouldn't have spread so quickly; and there'd have been no feedback loop to managers.

Coming out of all our research and thinking was the recognition that we fundamentally need to rethink how we run organisations and relate to our customers. The breakthrough point came in linking that social connectivity to customer value. Organisations have what you could term 'relationship capital', the sum of its interactions with a customer across multiple channels and touch-points. But when you deal with a customer, you don't just deal with the immediate individual, but also their network of family, friends and colleagues who also influence their purchase decisions. By managing connectivity, companies can improve their relationship capital.

We started off by throwing our net very widely: we thought of an individual's network in terms of his or her interactions with friends, suppliers, the local community, the government, and so on. But we soon realised we had to narrow this down to interactions between customers, between employees, and between customers and employees, and we had to be able to map these interactions – or connections. If you ask employees to map their networks, you never end up with anything that looks remotely like the official organisation chart, you get something much more complex. But we knew we'd have to tie this to concrete benefits: talking about mapping relationships at the moment gets you the kind of response from clients of 'that's very interesting: come back in three years.' While we're convinced that measuring relationship capital will become as standard in 20 years' time as valuing brands is today, we had to have a proposition which had immediate benefits. Our approach was now very clear: we could map relationship capital, and identify and implement ways in which it could be improved.

Or so we thought. The next watershed came when we road-tested the approach with a few contacts. It's a fantastic idea, they said, but clients want to buy implementable ideas, not consultancy. What you have to do is be able to say is: 'what you need is one of these' and 'we've got one and we can get it up and running in your business in two weeks'. The thinking we'd done provided an intellectual context, but we needed to overlay it with a more solutions-led approach. We needed a spearhead to begin the relationship: buy the package, and you may eventually 'buy' the idea of relationship capital.

We therefore decided to focus on service industries, and in particular on those sub-sectors which involve a high degree of customer interaction and where, as a consequence, the behaviour of employees has a significant impact. A good example is travel. If you've ever bought a skiing holiday, the chances are that you went with a group of people – almost everyone goes with family or friends. That means that, if you're trying to sell skiing holidays, it's not just the immediate purchaser you want to influence but that network. You've got to build a solution that appeals to everyone. So why not set up a web-page with details of the holiday on that everyone can look at? Why not build a relationship between the rep at the skiing resort and the customers (who normally don't meet until the latter actually arrive), by getting the rep to send text message snow reports to everyone in the weeks before they arrive? It's exactly the kind of thing that people will remember a year later, when it comes to book their next holiday. And it's not difficult to do, using technology.

Our selling proposition now focuses around building prototypes of ideas of this nature. We don't offer an end-to-end solution in terms of full-scale implementation, but we'll work with specialist technology companies to make it happen. We channel the innovative thinking into our 'context' and into the opportunities we identify, which will be unique to each client, but what we're offering clients is a way to improve the value of the customer relationships via some clearly defined tools. We're a loyalty factory, in effect.

Intellectual Capital: Towards a Value-Based Method

It's precision, not prescription, that consulting firms should be after:

- With a wealth of materials being produced, both by your own firm and your competitors, it's important to have a clear theme around which you can structure the services you offer. Conventionally, it's been the features of these services themselves or the areas of the consulting organisation that offer them that have provided a structure, but those are not things to which clients can easily relate. Grouping services round a theme with which clients can identify – connectivity, for instance – makes it easier for them to understand how what you're talking about fits with what they're thinking about.

- Test your ideas with clients. It's extraordinary how many consulting services are developed out of work with one client, but are then not market-tested with more. In converting a single client project into a product, you inevitably have to engage in a process of distilling what you did (the practice) into rules that can be applied more generally (the theory). For every subsequent client, you're then faced with the problem of proving that the theory has practical value for them. More 'prosumption', where clients and consultants work together to develop ideas, both at the theoretical and applied levels, will result in services that are better designed to meet client needs and can be described with clarity and detail.

- But 'themes' can be too broad: opportunities need to be created whereby clients can see how an idea may be relevant to them. It's crucial, therefore, to be able to describe practical instances of how your central idea can be applied – to specific situations in specific industries. This might involve limiting your market to those areas, or it might mean developing multiple applications, each individual tailored to different circumstances, but all joined by a common thread. Clients need to be able to fit the pieces of the jigsaw together for themselves, by relating the situations you describe to their particular circumstances.

- For clients who can see the applicability of your idea to their organisation, you need to provide concrete product that they can put their hands on, prototype, show their colleagues, and so on. In other words, you need to 'package' sub-sets of these applications into concrete processes or software. These packages, in turn, will provide low-key, non-threatening entry points into a potentially larger-scale consulting project. It's not a buy all or nothing approach, but keeps clients in control by allowing them to 'buy' only part of your services, rather than have to commit to anything more.

- None of the above is possible without considerable top-down effort on the part of the consulting organisation. All too often, product development is something that is pushed down the hierarchy as a firm grows. The result is a bottom-up approach which, while it may exploit the experience of consultants working at the client coal-face, lacks any degree of strategic intent. The 'winning' services are simply those in the right place at the right time – filling a gap in a marketing brochure, pushed by a particular partner; they're not necessarily the best services.

3

Market Intelligence: Avoiding Junk Mail Consultancy

'No matter how you slice it, not all clients are created equal – most people would agree with that', says Bob Borsch, who is responsible for Global Account Relationship Management at PricewaterhouseCoopers.

❝ We need to segment our clients: we can't treat all clients equivalently, so you have to decide which clients require and deserve the most resources and attention, and can be managed for the maximum value of the business.

PricewaterhouseCoopers has three segments. There are the 'strategic' accounts which are our most important global accounts; 'managed' accounts are those which deserve ongoing attention; and 'portfolio' accounts are what we call targets of opportunity – typically these are one-off accounts where we're working on a single project. We have around 100 strategic accounts worldwide selected on the basis of whether the client is an ongoing buyer of the kind of consulting services PwC can offer. Typically, the managed account set is closer to around 350 accounts and those are the ones that are very important clients to a specific territory or centre.

It's incumbent on the firm to manage strategic accounts in a special way: we assign a partner to the account – the lead relationship partner – who wakes up every morning thinking about that account and spends the bulk of their time working on that account, listening to that client, spending time with them and showing them the services we can

> *What clients want:*
> Consultants who know exactly what they need when, and can help articulate the issue in a disinterested fashion

> *What consultants want:*
> A steady stream of work enabling survival and, in good times, growth

> *Value-based consulting:*
> Supplementing conventional account management processes to make better use of 'pull' technology and establish more of an ongoing dialogue between the consultant and client

provide. The lead relationship partner is responsible for managing their account to its best and highest use in the business.

But we have to support the account management process. You have to start with the assumption that not every partner was born knowing how to manage an account. Some partners can do it straight away, others need coaching: if you draw a normal distribution curve, some 20 per cent of partners are probably really good at account management, another 60 per cent are okay and can be educated and there's probably around 20 per cent who don't have a clue and will never be good at it. We've therefore developed an account relationship management methodology, to ensure that everybody uses the approach to account management. That's important because if you're working with, say, Johnson & Johnson, who have operations all over the world, then you want to make sure everybody's behaving the same way. We've also built a worldwide information system so that we can see how these strategic accounts are performing. This is essential: if, for instance, you were the lead relationship partner for General Motors, and the firm's doing $10m worth of work in Germany and another $25m in Australia, how would you ever know how this account's performing? We have a global data warehouse which extracts information from territories all over the world and collates it into our 'client portal'. Authorised partners and managers can go and look at the performance of an account, and drill right down to the engagement level, to see what kind of work is being undertaken. You've also got to help build the incentives. So partners' personal plans are set up so that they record the fact that they're in charge of this account. They're responsible for the account's performance. Finally, there's what we call the 'plan on the page'. Partners are asked to sit down and plan annually how they're going to manage the account, to analyse the client's issues and problems and identify what kinds of services we should be offering and what resources we're going to apply. We estimate what the future revenue streams might be, not only for this coming year but over the next three years, because strategic accounts are about long-term relationships, not one-off sales pitches. If you added up all these account plans for a year, we'd get an estimate pretty close to 40–45 per cent of our total revenue.

But what's the value to clients? Obviously, every client is different, and the perceived value varies from country to country, influenced by the local culture. We also customise how we manage accounts to suit the needs of each organisation – so it's hard to generalise. Some clients are very comfortable using consultants: they treat you like a partner, and see you as part of the solution to their business problems. They're open about what their agenda is and then plan with you how they should prioritise their consulting expenditure and where you might fit into that picture. That's kind of a true partnership. These clients appreciate that good working relationships ensure a lower cost of delivery, lower risk of failure and better results. But there are other firms who put everything project out to competitive tender, irrespective of what your relationship is. They'll say 'well, that's great, but we're still going to invite three other companies to bid for it'. We have to manage those clients very differently, because, by definition, as soon as you have to bid for something in a competitive situation, you're probably going to have to knock off 20 or 30 per cent of the price. But that's a false economy because you have to provide less, and erodes trust – we won't invest as much in a client that does not appreciate our investment in them. How we manage an account does very much depend on the client's behavioural patterns: where clients do expect us to compete, we obviously can't afford to give as much to the account in terms of services and support. And there are people who don't want to know they're a strategic account – they're suspicious that we're just trying to sell them something.

For clients prepared to work in a genuine partnership with us, there are many benefits to being a strategic account. They have access to the attention of a full-time partner and they have what we call 'core resource teams'. A client may, for example, be doing a lot of SAP work, so we'll assign additional members to the account team who are full- time experts in this area. Clients therefore get access to a team of people who are the specialists in the area they're interested in. Strategic accounts have first call on our resources: if they've got a big issue and we need to start pulling people from other teams and other places, we'll do that for them. But resourcing is never a trivial matter. So, we're now in the process of assigning a top management liaison – what we call a 'top-to-top' relationship – very senior people within PwC who, on a periodic basis, will drop in, meet with the client's top management and listen to their issues. It's something that we see sparking opportunities that nobody had thought about, but also providing some degree of clout to get people specific resources they might not otherwise have got. Finally, we try to give them 'best value' pricing. This doesn't mean the cheapest price because we may have dedicated some of our top resources in the world to an account – people who are very valuable and comparatively expensive. But we'll attempt to find them the best price around the world and have a relatively uniform approach to pricing worldwide.

PricewaterhouseCoopers is a vast pool of talent worldwide: there's almost no problem for which you can't find somebody that knows something about it and can solve it. A lot of the time, I think clients don't fully appreciate how powerful it can be to work closely with an account partner who can give them access to all that.

But we could probably do the job of leveraging our resources better than we do. Part of that involves doing a better job of educating them about what the account partner can do, but technology is likely to play an increasingly important role. We had one client which put links to the PricewaterhouseCoopers' web-site on their intranet, so that employees could come directly to us, whatever their query was. That's a classic information technology connection where both sides end up working much more closely together. We could also learn from the example of our clients, some of whom handle all their links with suppliers electronically. We have clients who are asking us to submit our bills electronically, and they want to pay us electronically. That's something that businesses like ours haven't done historically, but any electronic synchronised link – whether it's payment information, web-site linkages, or collaborative chat rooms or other things, where we and our clients could exchange information – would forge bonds that would be very helpful. Furthermore, account management is an expensive proposition: although you'll never replace the human interaction component of the relationship, there are still all the administration links and processes which support this interaction. If you look at all the touch points involved, there's a lot of things that could be automated. American Hospital Supply was famous for going to all the hospitals in the US and putting a computer terminal inside the hospital linked to their purchasing web-site. All you had to do was turn it on and you accessed their entire catalogue of products and services, and ordered anything you wanted to. The idea's a little bit more accepted now, but a few years ago this represented a great strategic advantage. There'd be similar advantages to be gained now, I'm sure, if we really exploit the kind of technology now becoming available. We're still pretty embryonic on that – I think most companies are – but I think it's the next **,** generation of account management.

Corporate Relationship Management

Every industry has its equivalent of junk mail – and consultancy is no exception. Much consultancy marketing is, in reality, selling: we've developed such and such a service and would like to invite you along to breakfast seminar so that we can tell so much about it that you'll buy it. And – except in those rare moments when a new market suddenly takes off – it's a tremendously inefficient way of working. That truism of mass media advertising – 'we know that we waste 50 per cent of our money, we just don't know which 50 per cent' – if applied to consultancy, would probably reveal that 90 per cent of marketing expenditure was wasted. So why bother? Partly, because you want to build brand awareness, not simply among potential clients, but among potential recruits and, indeed, your existing employees. Partly because that's what almost every firm does (even small firms, within their own spheres of interest), and you have to keep up. But, primarily, because you're paranoid that, if you don't stay close to existing and future clients, you may simply not be in the right time and the right place when they need your help.

It was this paranoia that gave rise to the idea of account management in the consulting industry in the mid-1990s. Essentially a means of creating a dedicated sales force in all but name (something which the 'professional' ethos of consulting had historically precluded), account managers took on responsibility for walking the corridors of large corporations, keeping in touch with developments, getting to know people. This move was further fuelled by two recognitions. First, that another adage of consumption – that 20 per cent of your customers contribute 80 per cent of your sales – could also be applied to consultancy. Smaller clients, smaller projects tended to absorb a disproportionate amount of selling time – time that could be better invested with bigger clients. Second, that, by keeping its ear to the ground, the consulting firm would have better market intelligence, enabling it to manage the peaks and troughs of business with which the consulting industry has conventionally been plagued.

Since the turn of the century, the consulting industry has seen a bifurcation, between corporations which – on the one hand – are according preferred supplier status to a small number of (usually large) firms and – on the other – are reluctant to engage consultancies for the kind of large-scale assignments that characterised the 1990s. The first of these trends implies that the client is prepared to have a long-term relationship with a particular firm, the second the complete opposite. And account managers have been finding themselves at the heart of trying to reconcile them. Thus, the account manager's role has evolved from salesperson-*cum*-spy to that of a classic broker, matching the needs of the client with the skills and resources of his or her firm.

Talk to clients today and it's clear that these two divergent trends are set to continue. As one client put it, 'what we want is to ensure that we have access to world-class resources as and when we need them. We do want people who know about our organisation, we just don't see why we should invest our time in briefing every consulting firm under the sun, when we can concentrate our efforts – and their efforts – much more effectively.' If anything, clients see the size of projects shrinking further in the continued backlash against large scale IT implementations and Y2K work in particular. 'It's important for us', said one, 'to break what we want to do down into its constituent components: that way we can understand much more clearly what we're buying. Just because we have a small number of preferred suppliers doesn't mean that we're going to hand over large pieces of work. Quite the opposite, in fact: we need to see how all the pieces of the jigsaw fit together.'

From the consulting perspective, this poses a problem. Preferred supplier status inevitably requires a sizeable investment in account management, but if the individual projects which emerge from this process are many but small, then the consulting firm may find itself back in the unenviable position from which – ironically – account management was supposed to extricate it. That the large client behaves more like a host of small clients where the cost of sales involved is too great to be recouped from the projects themselves.

In this environment, the role of the account manager becomes critical. In the first place, they have to become much more effective, in terms both of identifying as many potential sales opportunities as possible and of staffing these projects with exactly the right people. More than anything else in the consulting industry in the last twenty years, the account management principle has given clients a sense of the consulting 'continuum' within their organisation: a good project becomes more likely to lead to more work, a bad project to less work. The account management process has, in effect, become the organisational memory for consulting, and account managers have a crucial role in ensuring client satisfaction. Five years ago, if a client wasn't satisfied, they'd blame the immediate consulting team; now, they can blame the account manager – the broker whose job it is to make the right match. And it's not just effectiveness that matters, but efficiency: from the viewpoint of the consulting firm, the account manager has to gather more information at lower cost, for the whole model to continue operating.

The conventional model of account management presupposed that the account manager would be looking after a small number of client contacts on a largely face-to-face basis. With the pressure mounting to cover more and more ground within the client organisation, this model needs to evolve into one that can handle many more points of contact at lower cost (see Figure 3.1). The danger is that it evolves into one of two other directions: by lowering the amount of money invested, there's a risk that you lose effectiveness as well and end up with far fewer contacts, but if you increase coverage without reducing the average cost of contact, then you may have taken the first step towards bankruptcy. Clients want effectiveness (someone with a good understanding of their business always available); consultants want this too, but not at any price.

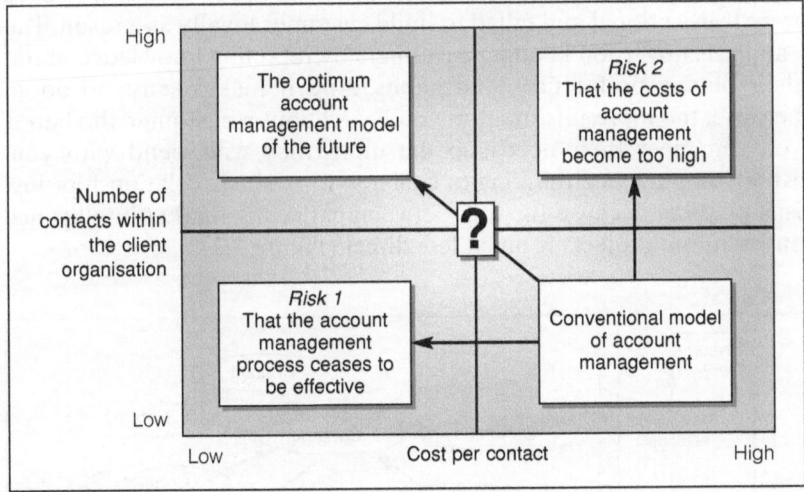

Figure 3.1 *The future evolution of account management*

In superficial terms, the choices the consulting industry now faces appear to be exactly those faced by almost other organisation: how best can you segment your customer base so that you concentrate your resources on those segments likely to produce the best return?

Being hard-headed, you could say that the only reason why it is worth cultivating contacts within a client organisation is to earn fees – either directly by getting to know key decision-makers, or indirectly, by getting to know people who will provide the intelligence you need to identify opportunities and decision-makers. The problem is that there's no necessary correlation between the two – in other words, you may get the best intelligence from the contacts with the least money.

Corporate clients are like individual consumers (indeed, what are they except the sum of a group of individual consumers?) in that they want consulting firms to approach them about issues in which they're interested at the appropriate time. But, from the standpoint of the consulting firm, the primary difference between *consumer* relationship management (in the business-to-consumer context) and *corporate* relationship management (in the business-to-business context) is that – for consumers – expenditure and information go hand in hand. One of the lessons of the dot.com boom was that it wasn't enough to have simple financial transactions with your customers: you needed to facilitate an exchange of information that – over time – would accumulate into a relationship. Too many web-sites were transactional and failed to build customer loyalty as a result. The emphasis now is on keeping customers by retaining knowledge about their buying preferences and habits. And it makes sense to do so because, the more information you have about a customer, the better you can meet their needs, so the more they will spend with you. Applying this same thinking to business-to-business relationships just isn't possible because the link between gathering market intelligence and winning projects is much less direct (Figure 3.2).

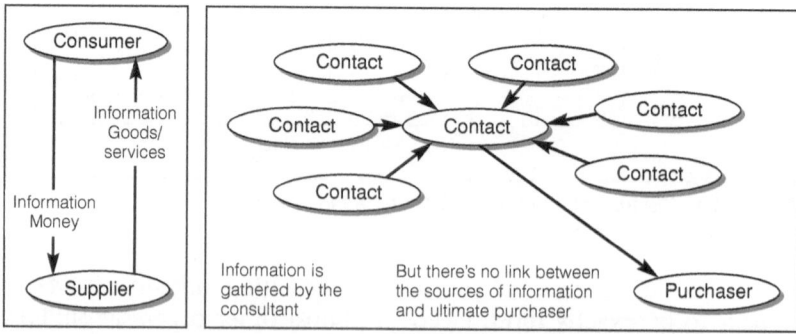

Figure 3.2 *The dissociation between information and purchasing in the corporate client context*

Finding out about a particular issue within a client's organisation may yield precisely the knowledge which enables you to win a contract, but the process of gathering data may have taken place in an entirely different area of the organisation. How often is it that someone you know well in one business unit of a large corporation points you in the direction of someone in another business unit? How many times has a relationship you have with one company in a sector provided you with useful insights into a competitor organisation? One-to-one marketing is predicated on the exchange of information for loyalty: the more information you have about people, the more effective (and profitable) your marketing will be, but how can this apply in an environment where the exchange of information and sales are separate activities? How, for example, will the client who has hired you appreciate that you won this work because you had access to better information from somewhere else in the organisation/industry? As far as they're concerned, you may just have been particularly good – or lucky – at guessing their needs?

Two things need to change. First, consultants have to persuade clients that telling them – the consultants - as much as the client can about a project makes for a better project for both parties. Clients win because projects are more precisely tailored to their requirements; consultants win because they know better what will be entailed and can plan and price accordingly. Some clients do this – but it takes a confident client – and a confident consultant – to do so. For clients, the temptation is to 'test' the consultants; there may also be concerns about giving confidential information away before the contract is signed and about revealing how big (or small) the budget is. For consultants, the fear is that they tell the client so much that they – the consultants – are no longer needed: the client can go off and do the work themselves. Second, the connections between the background intelligence gathered by a consulting firm and the quality of both its pitch for a project and the delivery of that project need to be made much more visible. In effect, more of a link needs to be made between 'tacit' knowledge – informal, undocumented information – and 'explicit' knowledge – the physical output of a consulting project. Without this connection, neither clients nor consultants will be able to appreciate the value of one to the other.

Client-Consultant Exchanges

Technology has to be part of the solution here.

Keeping in contact with a multitude of different people without the process costing the earth, aggregating all the individual elements of information that contribute to winning new work, making the link between that information and the quality of input a consultant can make may be better done by a computer than a person. Historically, the consulting industry has steered clear of such a solution, partly because the technology was unable to cope with the complexity of the problem, but mostly because there was – and is – a fear that technology would become a barrier in the client/consultant relationship. But as complexity and costs mount, and as the links between information and quality become more etiolated as a result, an increasing number of firms will pass a point of no return.

What will this mean in practice? Clearly, a crucial part will be to expand existing knowledge management – the majority of which is focused on capturing information once a project has been won and delivered – to cover the pre-sales period. But the problem here is that most such systems have been designed around formal, explicit knowledge – proposals, reports, and so on – rather than round the tacit, informal knowledge that is characteristic of the pre-sale period (Figure 3.3).

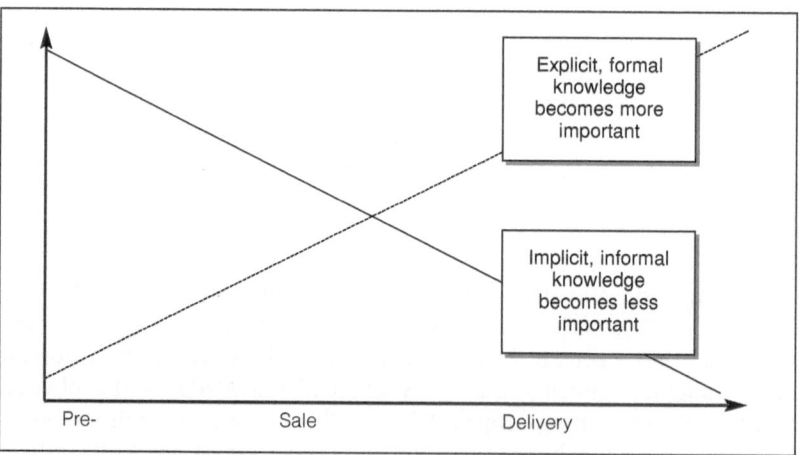

Figure 3.3 *The changing ratio of tacit to explicit knowledge in the life-cycle of a consulting project*

Firms whose knowledge management systems have a significant component of informal knowledge, which are also the firms that, historically, have tended to excel at managing relationships with corporate clients, will continue to have an advantage here, as their systems will be better designed to handle less structured data. We're also likely to see these more structured databases being complemented by 'self-organising' sources of information – virtual communities, e-mail chat rooms, where people involved with the same client/industry can record and exchange information. But the single most significant change that's required will be to switch from 'push' knowledge to 'pull knowledge'. Just as marketers have realised the need to switch from trying to interest consumers in what they have – push marketing – to getting consumers to tell them, the suppliers, what they want – pull marketing, so consulting firms need to switch from gathering information from comparatively passive client contact to encouraging those contacts to volunteer information on a much more proactive basis. But how can you encourage clients to do this? The idea of handing out loyalty cards or awarding air miles sounds laughable, but some incentive will be required. The obvious solution is knowledge: just as airlines have unused capacity which they can 'give away' to frequent flyers, consulting firms have 'unused' knowledge – information and ideas from projects or research which typically sit around gathering digital dust.

The whole idea of online consulting (which is discussed at greater length in Chapter 16) is predicated on selling this knowledge to clients who are unwilling or unable to pay a premium to have it delivered by actual consultants. For this reason, it's been largely treated as an 'economy service', earmarked for clients who are effectively travelling coach. But the same online databases may have a far more significant role to play in providing an incentive for even the largest corporate clients to investigate the issues that concern them. Capturing this information – who is worried about what – will provide consulting firms with an invaluable source of market intelligence and the clients win too, by gaining access to a database of consulting techniques. The worry has always been cannibalisation: that, by giving clients wholesale access to a firm's knowledge base, it will be giving them the tools to consult to themselves – it will cut out the middle man. But this will only happen – as it has already happened in so many sectors – where the middle man adds no value. Online consulting will only cannibalise a firm in the process if that firm does not customise its essential service offering to the needs of individual clients.

Spending to Save in the Client Relationship

But gathering intelligence is only part of the picture: it's equally important to be efficient in the way you interpret that information and develop solutions to the problems that emerge.

' Keeping the innovation flame burning during periods of economic slowdown isn't easy', acknowledges Randall McComas, Scient's General Manager for Business Development. 'The key change we've seen in the current climate is that, when we go in for an initial, introductory meeting with a client, we must take more intellectual property with us than we've ever had to before.

That's because clients are trying to figure out what to do next, now that they've already done what the textbooks have told them to: cut back on discretionary expenditures, reduce overheads, and so on. Now they realize they must also find ways of bolstering revenue. They're looking for new ideas, but they're watching every dollar they spend on them. Scient is now finding success at conducting the detailed level of due diligence before we've even met with the client that we might, in the past, have done at the start of a consulting project. Now, when we go into a potential client that we haven't met before, armed with a thick book of analysis about their business and industry, with suggestions about opportunities for raising margins, more often than not, they're telling us they've never seen anything like it. After presenting the 'book' we then dive right in with a series of options – already researched – to discuss, and we are therefore absolutely unequivocal about how we'll add value. We're able to start resolving client's needs from day one of a project, meeting their thirst for immediate gratification. It's something that clients have learned they need to do for their own customers, and it's what they also want from their suppliers. Essentially, we're each applying the same customer relationship management techniques – clients to their customers, Scient to its clients.

We also think it gives us a unique competitive edge. We differentiate ourselves by seeing a company through the eyes of its customers. How can we improve a customer's experience with the company? What aspects of the relationship are of greatest value? Will there be any customer benefit to a service being delivered on-line? What is the desired experience your customers, partners and employees want? How can you provide it? When you match a desired experience with proven process you learn how to create a company that is focused on meeting the needs of its constituents and driving true value. This is where competitive value is found. It can only be achieved by companies that can help their clients not only understand, but implement effective 'experience management' strategies.

Can Scient do this on a sustainable basis? We believe strongly that we can. It's a knee-jerk reaction to see expenditures like these as having a negative impact on a firm's bottom line – the thinking that, if we've invested 20 days preparing the document, that is 20 days that has to come out of our pockets. But what we're finding in practice is that it actually reduces our cost of sales, by cutting out a whole host of the 'to-ing and fro-ing' – the courtship that has conventionally preceded consulting engagements. Clients want to cut to the heart of the matter as quickly as possible – and we're putting together content and processes which help them do just that.

Everyone wants out of this recessionary environment: they want innovation – but it has to be applied innovation, aimed at extracting better value from their existing assets. They don't want to buy a new 'vision': they want to buy input from people who can help them with the practical means of implementing the vision they already have. 'Applied' strategy is something that Scient has considerable experience in. From our previous work, we have a reservoir of relevant data – customer research, usage statistics, channel switching data, and so on – which we can analyse and reapply to a client's unique circumstances.

We've also developed very efficient processes for bringing experts from different fields together in a 'war-room' for a limited period of time to develop these client packs. Our key objective is to come up with an analysis, based on the data we've amassed, of an organisation's pain-points, and then generate a lot of innovative thinking around possible solutions. And we're extremely detailed so that we can present clients with concrete solutions, not theory. Scient's merger with iXL, which was completed in November 2001, has provided complementary skills that we can feed into this equation. iXL's breadth of experience was huge, and they brought targeted solutions to the merger. Scient had a track record in packaging strategic thinking into practical solutions. Bringing the toolsets together means that we can think broadly, yet focus narrowly – and that's perfect for this environment.

Both Scient and iXL also had, independently, very collaborative environments which they bring to the newly-merged company. When something is going on, our first instinct is to roll up our sleeves and jump in. That culture has been fundamental to our ability to bring people with different skills – strategy, marketing and technology together, often at very short notice. We pull together people who can generate ideas that cut across conventional ways of thinking. But we also believe in a healthy amount of peer pressure: we circulate documents between these conceptual war-rooms so that people can see the standard of what's being achieved by other groups and work to match, if not exceed, it. It helps as well that we're good at celebrating achievements: we're not the kind of organisation that picks holes in something just because we didn't create it. We're much more likely to say 'hey, man, that's fantastic' and try and use it in our own work, rather than dismiss it.

I can't overstate how important I think the cultural side is. Finding ways of being innovative within Scient without spending insane amounts of cash is exactly the kind of thing we're looking to do with clients. That sends out an important message **,** to our clients and reinforces our brand.

Market Intelligence: Towards a Value-Based Approach

To ensure that they turn up on a client's door at precisely the moment they're needed with precisely the right idea, is something that would benefit both clients and consultants. Clients would benefit because they'd have people coming with solutions to real problems, rather than a seemingly endless stream of brochures, books, meetings and presentations which may or may not be relevant to their day-to-day concerns. Consultants would benefit because they'd be able to cut their cost of sales substantially – providing that the costs of gathering and analysing the information were not prohibitive. Being able to deliver this value to both sides, while minimising the costs of doing so, requires two things – visibility and collaborative working.

The problem with having dedicated client partners, whose job it is to 'walk the corridors' talking to people, finding out what's going on, it that it's a continuous process of pulling information from clients, expensive because the relevant information has to be hunted down. It also sounds, when people talk about it, slightly clandestine, as though a benevolent corporate spy had been allowed to infiltrate the client's organisation. Essentially, it's expensive – and potentially invasive – because people don't always know it's going on. It's true that an account partner will probably have an opposite number in the client's organisation, a relationship that varies in its formality. For a corporation operating a preferred supplier list, the role may be explicitly designated, required for contractual reasons. For another organisation, which has worked with a consulting firm over a long period of time, the role may simply have evolved out of personal relationships. Either way, the knowledge that the client partner from the consulting firm is 'available' for informal discussions may be very limited within a client's organisation. As a result, there are many occasions when a client partner feels inadvertently by-passed, because someone in 'his' or 'her' client has approached another consulting colleague first. The solution to this is likely to be technical: joint intranets which both make the consultancy's presence obvious to everyone and facilitate the exchange of information.

But collaboration between client and consultant, in terms of identifying possible problems, has to be mirrored by greater collaboration with the consulting firm itself. The 'junk' consulting sales pitches don't happen simply because the level of understanding about a client's real needs is poor; they also occur because little effort is made to make best use of that intelligence – and return to the client with customised solutions – except where the scale of a project is very large. As the average length of projects shrinks, consulting firms will need to be able to deliver this customised thought process much more efficiently, or risk losing the sale to someone who can. And the key to doing this lies, not so much in software, but in the ability to bring people from different backgrounds together for very short periods of time, and use this intensive environment to spark new solutions to which clients can instantly relate.

4

Brand Consistency: Delivering Experience As Well As Services

Clients don't believe that a single firm can do everything. 'That's not credible', they're saying. 'What do you really do?' It's a bit like a Christmas tree – underneath all those glittering trimmings, there's a solid tree – and that's what clients want to see – but some trees are so heavily decorated that it's not easy to look beneath the surface.

Daniel Flamberg, Senior Vice President, Digitas, Europe

Seeing things from the client's perspective, brand is a substitute for accreditation. It's what clients turn to for reassurance that the consultants they're hiring will at least 'make the grade'. The more standard the project, the less important brand will be as part of the purchase equation (which is why the brands of auditing companies are significantly less well developed than those of consulting firms). The more innovative and individualistic it is, the more important brand is. Why? Because a firm that undertakes standardised work – by which I mean projects governed by a formalised methodology – can:

What clients want:
A guarantee (of sorts) that what they see is what they get

Value-based consulting:
Creating and living a brand capable of reinventing itself in response to changing client needs

What consultants want:
Control their exposure to the potentially variable performance of individual consultants

- Train (and accredit) their own consultants in the methodology;

- Acquire accreditation from an outside supplier (for example, a software house);

- Demonstrate a track record of similar work to prove their credentials.

A firm working in more innovative, fluid areas can do none of these things: no methodology exists in which to train their consultants; no third party can accredit them (except perhaps in some supporting tools); no parallel work exists.

Moreover, the inverse relationship between brand and accreditation is a dynamic one (Figure 4.1). The more a firm standardises its services, the less important its brand will be. Conversely, if a firm that has been operating in a standard market moves into more innovative work, it doesn't just have to change its brand, but invent a substantially more powerful one. But a brand is something more to a consulting firm: it's also the means by which a firm – large or small – contains the power of the individuals working within it. If the personal brand of an individual consultant is stronger than that of a firm, then he or she can leave the firm confident that their clients will follow them. But if the firm's brand is the stronger, then the individual is primarily dependent on the firm for work. Again, this is a dynamic relationship: a consult who's graduating into full-blown management guru will eventually leave his or her firm. Indeed, the firm's brand may not be able to contain the consultant's personal brand: like the combatants in a western, there's only room for one brand in one town. Conversely, when times are hard or when the market's attention has moved on, one-time gurus may find themselves retreating under the corporate umbrella.

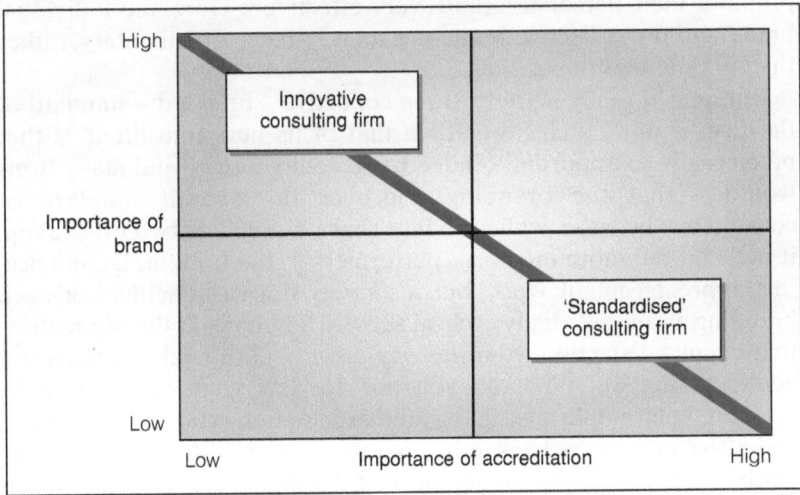

Figure 4.1 *The inverse relationship between brand and accreditation*

Of course, the stronger a corporate brand, the longer it can absorb an individual's growing personal brand. Brand is, ultimately, a secession of power from the people who make up a consulting firm to the firm itself, and it's therefore no coincidence that consulting brands first appeared in firms that already had a strong 'one firm' culture and that they've been much slower to take hold in firms where authority is still distributed across a multitude of comparatively autonomous partners and business units.

These different functions of the brand bring clients and consultants into potential conflict. For clients, a brand is a means to an end – a way of guaranteeing a high standard of service. This is also true for consultants: the problem is that brand – or the subjugation of the individual to the rule of the corporate entity which it involves – is also an end in its own right. Forced to choose between the two, what does a consulting firm decide? Sacrifice its brand on the altar of the individual excellence and a major part of its collective *raison d'être* evaporates. Rein in the would-be gurus, and it runs the risk of dumbing down its entire organisation. Thus, while clients may get a consistent standard of delivery, it may not be to the highest standard that individual consultants within the firm are capable of achieving.

It's interesting to note that brand among business schools is managed very differently. Part of the aim of a business school is to produce management gurus – to pick the future winners and ensure that they have not only the skills but the channels to market that promote their personal 'brand' very effectively. Here, the individual brand and the collective brand are seen to be complementary, rather than in potential opposition.

But does it really matter? If the collective – branded – standard of the firm is one notch lower than that of its best consultant, is that notch really so important? Indeed, you could argue – and many firms would – that the branding improves the overall standard of consultancy because, while it might shave a notch or two off the top, it pulls up the more mediocre performers at the bottom. Done once, this approach might work, but it's surely not sustainable. Let's see branding as an essentially cyclical activity. Someone at the top realises that client satisfaction is on the way down – nothing dramatic, just a gentle tailing off. What do you do? Review your brand strategy; measure your employees against the expectations your brand creates; bring the mavericks back into line; shed the worst performers and work to improve the skills of many others. As a result, client satisfaction improves. You maintain your strategy but, over time, other issues come along to distract you and to absorb your limited resources. Before you know it, client satisfaction is on the wane, and you start all over again. Without any radical rethink, it's likely that this process produces a slightly lower standard each time. Yes, you lose the worst performers, but you also lose talented individuals from the top. With fewer good people at the top, it becomes much harder to create 'stretch' goals – there are fewer inspirational role models to follow. The overall standard falls. The brand itself starts to look tired and unconvincing: more good people leave (Figure 4.2).

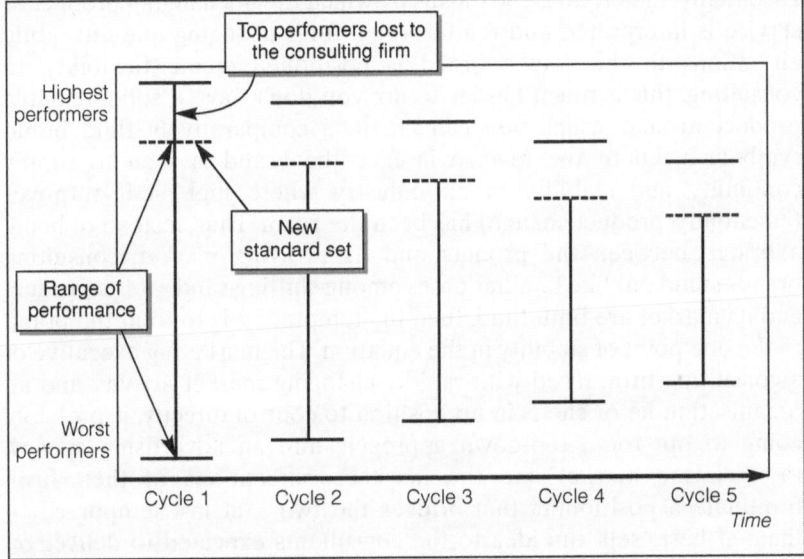

Figure 4.2 *The brand trade-off – sacrificing quality to brand consistency*

Brand Elasticity in the Consulting Industry

But do quality and consistency have to be in such relentless opposition? What can a firm do to create a brand experience which genuinely raises levels for all involved? And how can it do this without losing talented people who don't want to subsume their own brand into that of the organisation as a whole?

The whole idea of a brand as something that has enduring power comes from the consumer brands sector where it's possible to re-position brands by subtly adjusting the meaning the brand has for subsequent generations, leaving the physical product (and, usually, its physical trademarks) almost unchanged.

> A power brand is characterised by the distinctive nature of its brand personality, by the appeal and relevance of its image, by the consistency of its communication, by the integrity of its identity and by the fact that it has stood the test of time. But, of course, power brands have to evolve so as to remain contemporary for each new generation of consumers. What appeals to the young of today may be very different from that which appealed to the young of the 1920s, 1940s or 1960s. [1]

Essentially, the brand is the means by which the unchanging product or service is interpreted and reinterpreted for a changing market: subtle alterations to the way a brand is positioned create flexibility. In consulting, this is much harder to do: you don't have a solid, tangible product around which you can create a comparatively fluid brand symbolism. Quite the reverse in fact: the brand is seen to supply continuity and stability in an industry where high staff turnover (essentially, product change) has been the norm. Thus, instead of being a bridge between the product and its evolving market, consulting brands stand out like familiar oases among shifting sands; if the product and its market are both fluid, then the temptation is to treat the brand as the one point of stability in the equation. The marketing executive of a consulting firm, faced with rapidly changing market activity and an organisation he or she is in no position to control directly, is probably going to opt for a top-down approach: hire an advertising agency, research the market and the market's perceptions of their firm, formulate a positioning that bridges the two – at last temporarily – then, at best, 'sell' the idea to the consultants expected to deliver on their promises, at worst, try to impose it. Brands in the consulting industry are therefore more likely to become rigid (everything else is changing around them) and out of step with either the product or the market, or both. One person I interviewed summed the situation up: 'as director of marketing you're often outside the mainstream consulting practice and you don't have the power-base that practising consultants have. It used to be hard just to get people to talk about marketing: that's changed now, but it's almost as though we've gone to the opposite extreme – from decentralised marketing to absolute command and control. We spend a lot of money, hire a top agency and that's it. Obviously, we let the consultants know what the adverts are going to look like, but there's very little attempt to say "and this is what it means for you". We're making a big "if we build it, they will come" assumption.'

In their book, *Creative Destruction*, Richard Foster and Sarah Kaplan argue that organisations need to build themselves around discontinuity:

> 'Cultural lock-in' – the inability to change the corporate culture even in the face of clear market threats – explains why corporations find it difficult to respond to the messages of the marketplace. Cultural lock-in results from the gradual stiffening of the invisible architecture of the corporation and the ossification of its decision-making abilities, control systems, and mental models. It dampens a company's ability to innovate or to shed operations with a less exciting future. Moreover, it signals the corporation's inexorable decline into inferior performance.[2]

So far as the consulting industry is concerned, Foster and Kaplan could have added brand to the list. The more successful a brand is, the harder it will be to change but the more important it will be to change, if not now, then at some optimum point in the future before the brand begins to decline. The implication is that consulting firms will have to switch from trying to build lasting brands to creating disposable brands – brands which (unlike conventional consumer brands) have a solid, unchanging link to their product but a short shelf-life. Periodically, firms will have to reinvent what they stand for, redefining themselves in such a way that they can continue to accommodate the gurus emerging in their midst without either side compromising their identity.

Being able to create brands which reconcile the needs of both clients and consultants will therefore depend on being able to build a more flexible brand, one that can move dynamically, in relation to both a changing market and a changing product. And the key to this has to lie in creating a brand that is indivisible from the people who deliver it.

Reinventing Andersen's business consulting practice

Ashley Unwin heads up Andersen's (formerly Arthur Andersen's) People and Change Practice: most of his time is spent with brand-name corporations, working to ensure that their internal organisations can deliver on the promises made in external marketing. Following the firm's separation from Andersen Consulting (now Accenture) and its own rebranding campaign, it's an issue which is as pertinent to

Andersen as it is to its blue-chip clients. Unwin therefore played an important role, not only in developing the 'new' firm's strategy, but also in ensuring that it could be successfully implemented across the organisation.

' Most strategies come out of an opportunity. You see a gap in the market and you go for it – essentially it's a linear process. We came at Andersen's strategy from three angles. First was the gap itself: the consulting industry was – indeed, is – polarising between the niche firms and the large-scale systems integrators, but, in the wider context of professional services (tax and audit, as well as consultancy), clients will still want a predominate supplier of services who can either do everything itself or be an aggregator of services it cannot provide. The second dimension was our capabilities: we had to understand what we were best at, if we were to develop a strategy which was in any way realistic. Finally, there were things we wanted to do with the firm – our aspirations.

From the strategy that emerged from these discussions, we formulated a statement of the capabilities we would need to build if we were to do what we wanted to do. And we tried to engage as many people as possible in this process: the debate about our values alone involved more than 2,500 people. We were building a vision of what we wanted to do and who we wanted to do it with; we wanted to see how people would describe the cultural characteristics of the type of organisation capable of delivering on this vision. This exercise yielded around twelve characteristics which would drive the behaviour of the individuals within the organisation – being passionate about clients, for example – and from this we could, first, define the core competencies we already had and those we still needed, and, second, the rules the organisation would require if the values we envisaged were to be nurtured. We carried out value 'audits' in every practice across the world, benchmarking where the practice was in terms of its ability to deliver the firm's vision. In doing this, we had to take account of the different markets different practices were working in – Andersen in the US is very different to Andersen in the UK; in Turkey, we're rated the number one employer; in Poland, soap opera heroes work at Andersen.

Coming full circle, we mapped the results of this back to our brand attributes, to give us our value proposition for clients, a cultural blueprint for the organisation as a whole and a definition of the performance measures that would keep the two things in synch.

One of the issues for the consulting division was how to reconcile some of the values to a wide variety of working environments. We want people to be innovative, but it's difficult to be that if you're working on one of our larger, more complex projects: there's an inevitable tension between generating new ideas and maintaining the structure and discipline demanded by some projects. If we couldn't measure people's actual innovation here, then we'd at least measure their 'propensity to innovate'. We had to recognise that the people most likely to tell us what the markets would want in the future were our youngest, most junior people – people who were already using new technologies to stay in touch with their friends, listen to music, and so on. We therefore had to encourage these people to make more of a contribution – we launched ideas-generating programmes and Fish Tank, a project in which new graduates could get together to discuss the latest trends. By and large, we relied on informal rather than formal structures – peer pressure in effect – to promote innovative thinking.

We also identified some common themes across consulting clients – market integration would be one example – and we asked everyone to think about how this issue would affect the client they were working for at the time. Another initiative we took was to try and break down the internal barriers in our organisation, particularly between the strategists and technologists, because we saw this as being absolutely essential to our being able to deliver end-to-end consulting in specific areas. So we challenged everyone – irrespective of where they were working in the value chain – to articulate the intellectual property they brought to a client engagement. We created 'hot houses' to enable people to explore some of those ideas in greater detail. Once again, these were often the most junior people in our organisation; where their ideas merited it, we'd arrange a meeting with senior people on the client's side, giving the consultant a chance to argue their case. Several of those are now on the desks of the chief executives – something that's changed the perception of Andersen externally and internally. Clients can see that we're generating new ideas, and our people can see that we genuinely want to promote and reward innovative behaviour. We've succeeded in linking one of the brand values we aspire to – innovation – to our culture by tasking everyone with thinking it through in relation to a real client scenario: that's good for morale and client retention.

This is something we've been doing for clients, especially in the media industries, and now we're applying the same philosophy to ourselves. Essentially, the process comes down to unbundling an organisation's assets from its organisation and culture. How does a company's brand link to what it's doing? This is an incredibly important question in a world in which information travels so quickly – you've only got to look at the problems that some major brand corporations are having over using child labour in developing countries to see that. In the battle for customers, you've got to be able to deliver very slickly on your marketing promises. You can't be too prescriptive: you can't legislate to make people behave in a certain way at the most detailed level. What you can do is establish some pretty explicit rules and make sure that everything in your organisation is aligned so as to encourage the behaviour you're looking for. You have to start at the very beginning – when you recruit people – and be absolutely consistent from then on. You also have to be clear about what you don't do – the activities which, if you attempted them, you couldn't deliver on.

Peer pressure and establishing the right role models is critical: unlike the military, most organisations don't have the power to punish, so there's little point in coercion. Ours are essentially civilised organisations where the rights of individuals must be respected. And it's crucial that the most senior people 'walk the talk' themselves: if they don't, everyone will soon get the message that lip-service is enough. You have to stand by what you're asking others to do.

Does this mean that we will become too opportunistic, too *ad hoc*? We are exerting influence here, albeit indirectly: we're encouraging a particular working style, rather than telling people what to do; we're giving them rules for deciding between right and wrong, not deciding for them. Role models are also highly flexible: we don't have to waste time creating a rigid definition of the behaviour we're looking for, only to discover that the market has moved on. Role models are one of the most important ways in which we ensure that our brand values and organisational culture can adapt to change. They also mean that we can promote innovation as a brand value, without being too prescriptive – which would clearly defeat the object. We don't expect our role models to toe a party line: we expect them to work out what they should do in any given situation – people don't add value in identical ways and we don't want a culture that assumes they do.

It's been said that organisations judge themselves by their intentions, but are judged by others according to their actions. What we've been trying to do is try to reconcile the two, **,** at both the organisational and individual level.

Brand Consistency: Towards a Value-Based Approach

The disposable brand: it sounds like a contradiction in terms. But what is the rationale for adopting conventional ideas about brand that have primarily emerged from the manufacturing sector? A product can be enduring in a way in which a consulting service should not be: chocolate survives because people continue to like it, but customer relationship management will endure only as long as there's a perceived business need for it. Similarly, why adopt the manufacturing sector's top-down approach to branding, when the output of a consulting firm is the work of its individual consultants.

If a brand is to be something more than a strategic veneer in the consulting industry, it has to be generated from all sections of a consulting organisation, and managed from the top. What happens in many cases at the moment is almost the opposite: the brand is generated at the top, and then 'managed' at the bottom, in the sense that the consultants do or don't act in support of it. But there also has to be the crucial recognition that creating a brand isn't a one-off exercise. As people come and go from the organisation, as clients' needs change, a firm's brand needs to be continually recreated; that is, if it's to be – from the client's viewpoint – a way of reflecting how consultants from a particular firm will behave in practice, and – from the consultant's viewpoint – something more than a means of controlling the people who don't fit in.

---

[1] Donald Keough, former President and Chief Operating Officer, The Coca-Cola Company, 'The Importance of Brand Power', in *Brand Power*, edited by Paul Stobart (Macmillan, 1994).

[2] Richard Foster and Sarah Kaplan, *Creative Destruction* (Financial Times Prentice Hall, 2001).

Part 2
The Right Project

5

Specialisation: Letting Market Forces Prevail

People businesses are the hardest businesses to run, although in many ways they're the most rewarding. I sometimes say that if I could come back in another life, I'd like to have at least a few physical assets as well – give me a bit of machinery or something!

Adam Gutstein, Chairman, DiamondCluster

Specialisation and integrated working (the subject of the next chapter) are mirror images of the same issue. They both stem from clients' almost universal – and completely comprehensible – desire to hire experts. 'Ten years ago', said one client I talked to, 'people would have listened if you said you were a business consultant. You might have known a fair amount about a particular subject, but you'd also be expected to know about business "in general". I don't believe business "in general" exists anymore – and neither should "business consultants". 'It's not breadth we need', said another client, 'it's depth'.

> *What clients want:*
> To have access to world-class expertise in specific areas

> *Value-based consulting:*
> Reducing the intervention of the corporate firm, to enable individual consultants to develop and switch specialisations

> *What consultants want:*
> To retain a flexible organisational structure, capable of responding to changes in clients' needs

It sounds like a simple issue: clients want more, more specialised consultants; consulting firms have to redesign themselves to deliver this. But, scratch below the surface, and a rather more complicated picture emerges. First, there's the behaviour of clients themselves. The same interviewee quoted above also eulogised about the role that strategy consultants had played in her company. Indeed, talking to clients over the last couple of years, one of the areas where they've been most positive about the contribution that clients have made has been in developing business opportunities – something where having a strategic perspective has been critical. A second factor that complicates this issue lies in the commercial constraints facing consulting firms: the more specialist a firm becomes, the more exposed it is to changing market conditions and the less flexible it is in terms of responding.

When it comes to specialisation *versus* generalisation, you could argue that consulting firms fall into one of three categories, each of which face particular challenges in meeting the needs of their clients. The first, most obvious category is of the niche specialist – firms that are usually, though not exclusively, small in size and highly focused on a particular market. The challenge here is how to grow: like any nascent business, it's tempting to stray off your home territory, partly to have a more diversified portfolio so that you are less exposed to downturns in your niche, but partly because the growth figures in other sectors may be very attractive. A degree of hubris also seems to be part of the picture. As one former director of an e-business consultancy observed, 'you get to a point where you've grown so quickly that you think you can do anything. So you try to do everything – and you think you can't fail, that you're not subject to the same rules as everyone else.' The challenge here, then, is about being able to grow and manage your exposure, but without diluting your core specialisation, either in terms of your knowledge base or your positioning *vis à vis* the market. At the other end of the spectrum are the strategy consultancies, where too great a level of specialisation may mean that people no longer have a sufficiently wide view of the market to be able to identify the kind of business opportunities which clients rate so highly. 'It helps if you've got a unique idea,' commented one strategy consultant I talked to in late 2000, 'but this is becoming less and less common, and in any case doesn't guarantee success. To have any chance of being sustainable your idea has to address a specific "pain point" – such as an inefficiency or lack of information in a particular sector. Without this, and without a sound business model in which you've thought through where your money will come from, you've very little chance now of getting off the ground'. 'Using consultants is very expensive', acknowledges another, 'you can employ a lot of full-time people for the same amount of money. So we have to be sure that using consultants gives clients an edge. How? Because we've done it before, and our experience helps a client get to the right answer and get to it more quickly than they would have done without our help. When a company asks you whether its web-based businesses are achieving their full potential, they're posing real business questions. What are the prospects for these businesses? Where are their sources of profit? You need years of experience if you're going to be able to answer these types of questions quickly and precisely. Why get a technical specialist firm to do this, when their background will probably have been in advertising and/or technology?'

In a sense, both these extreme points on the spectrum are equally specialised – it's just that the nature of their specialisation differs. For the niche player, specialisation is in a particular technical area – a specific business process, or software; for the strategy firm, specialisation is in the process of formulating strategy.

The third category of firm – 'generalist' consulting firms – falls somewhere between these two stools: too technically-inclined to be strategic; too strategically-inclined to have developed any technical depth. Unlike the other two categories, the skill and knowledge base of generalist firms is typically out of kilter with their market positioning. A niche firm delivers niche skills and knowledge, and that's how it markets itself to clients. Similarly, a strategy firm markets itself as a strategy firm. But, all too often, a generalist firm finds itself marketing strategically-inclined skills in an area requiring deep technical expertise, or marketing technically-inclined skills where clients are really looking for strategic advice (Figure 5.1).

Of course, firms evolve over time. The generalist category is populated by niche firms that sought to expand and diversify, but ended up diluting both their skill base and their brand, and by strategy companies that either wanted to offer their clients an end-to-end solution and/or move into the apparently more lucrative delivery markets.

And, for those firms – the 'generalists' – caught in the middle, the problems are almost intractable.

The whole notion of core competencies has encouraged us to view organisations as portfolios of specialist resources, whether these are products, people or processes. No corporation pretends to know everything: so why should a consultant? Yet I'm continually amazed, on the courses I teach in business schools, how many people want to go into consultancy because they now – courtesy of their MBA – have a sufficiently broad view of business trends to be a credible 'business' consultant. Consulting clients' desire for specialist skills and knowledge (technical and strategic) remains constant (if anything, it's been

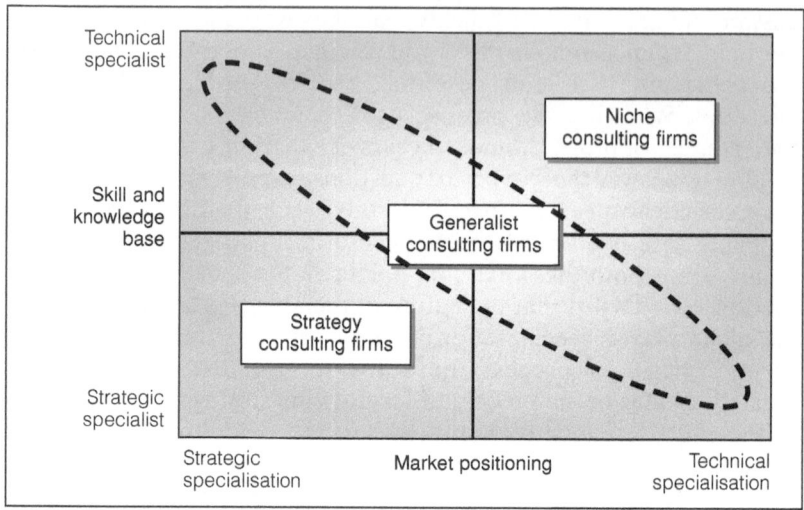

Figure 5.1 *Comparing the specialised skills and knowledge base of consulting firms with their market positioning*

growing), while generalist firms appear to go through repeated cycles – from specialism to generalism, from generalism to specialism. Every few years, the gap between what clients want and what these firms provide becomes sufficiently great to allow a whole army of more specialist firms to enter the market. Over time, the difference between incumbents and new entrants erodes: the incumbents 'correct' the direction of their organisation and bring it more into line with clients' needs; the new firms, buoyed with their initial success, lose focus and dilute their core speciality with a host of other skills and services. But just as the overall level of specialist skills within the industry rises, clients move on, finding new management tools and techniques. Skills which were valued become redundant. The consulting industry has been effectively – and instantly – deskilled. And because the consulting industry continually has to correct its course, the overall gap between the level of specialisation required by clients and that provided by consulting firms continues to widen (see Figure 5.2).

As clients' sights switch to new targets, it becomes easy to equate 'old' skills with generalism, if only because the 'old' inevitably lacks the apparent specialisation of the 'new'. It's a situation that's compounded by many consultants themselves. Faced with a client who's now interested in something else, it's tempting for a consultant to dredge up

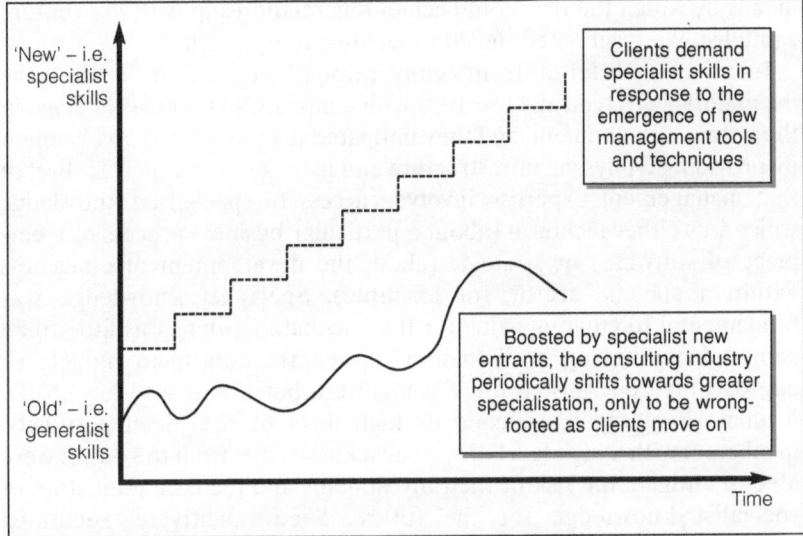

Figure 5.2 *The specialisation 'gap' between clients and consulting firms*

whatever knowledge he or she has on the new hot topic: it's a case of a little learning being a dangerous thing, and is transparently obvious to the client concerned. Thus, at a blow, the consultant is consigned to the generalist heap.

Nowhere has this more plainly been shown than in the now largely defunct incubator industry. 'The incubator idea sounded great on paper', said one (surviving) entrepreneur, 'but we quickly found that incubators – at least those we spoke to – really didn't know that much about business. They took a surprisingly long time to take a decision and we ended up feeling that we'd be teaching them not *vice versa*.' We eventually signed a deal with a Big Five firm in which they'd help us approach the investor community on a fee for results basis. But what we found was that, although the brand leverage of being associated with such a firm was useful, they didn't really understand the market and where it was moving. They had text book solutions rather that a fast moving dynamic approach. They provided some help in formulating our business plan, although none of it was exactly rocket science, and they were helpful in finding us a marketing agency. But the real disappointment came when they drew up their short-list of potential investors. There were only the obvious candidates, and when we went along to the meeting it was dreadful – it was clear that we were just a means by which the firm could cement its relationship with the venture capitalist. We terminated the contract after four months.'

Incubation differed from 'conventional' consulting in that the incubators were typically involved with companies at an earlier stage in the latter's development, and they mitigated the risk of that involvement by providing a physical infrastructure and management expertise. Part of that management expertise involved access to specialised knowledge which was either technical (about a particular business process or a new piece of software) or strategic (about the development of e-business within a specific sector, for example). Specialist knowledge was fundamental to creating value for the 'incubatee' (there was little to be gained from having a discussion of generic e-business models or approaching generalist venture capitalists), but it was also key to the incubator's ability to mitigate its high level of risk. Seen positively, incubators with a high level of specialist knowledge from the outset were able to mitigate the risk of their investments and increase their store of specialist knowledge for the future. Seen negatively, generalist incubators will find themselves unable both to reduce their level of risk and to learn for future experience (Figure 5.3). Despite this, too many incubators focused on offering general business advice at a stage when their clients needed this least.

Bain & Company is one of the leading global strategy consulting firms that tried hardest to grapple with this conundrum – balancing an overall framework with a need to keep testing it against the market, the collective firm with the individuals within it.

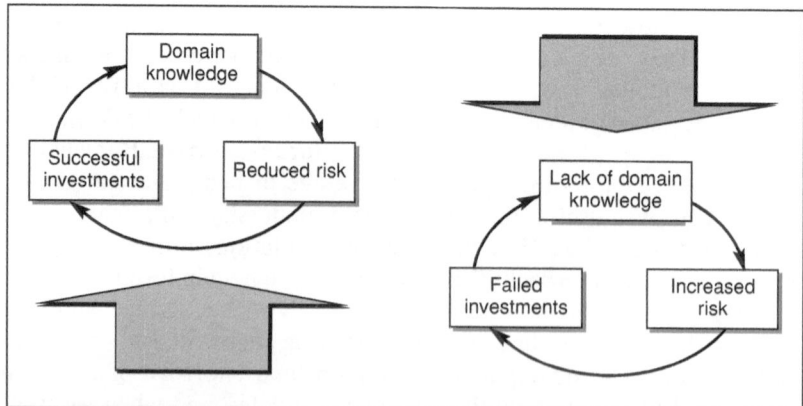

Figure 5.3 *The virtuous and vicious circles of incubation*

' Career 'structure' at Bain is a bit of an oxymoron', says Steven Tallman, Bain's Vice President of Global Services, and responsible for looking after the firm's training, knowledge management and information technology. 'We have an absolutely entrepreneurial culture: we believe that a higher proportion of people, when they leave us, go to start-up companies. We have always valued people prepared to step out on their own, and having a 'non-traditional' career is not something that breaks our rules, so much as reinforces our core values: we ask people to be entrepreneurial with their own development – as well as our clients.

Part of that is self-interest. Irrespective of downturns and upturns, the best people are always hard to find, and, once we have recruited them, we think that keeping their lives interesting is one of the most effective ways we can retain them. Being able to provide people with a wide range of experience is also an important way in which we ensure that our people have the breadth of knowledge and experience they – and our clients – require if they are to be credible in undertaking the type of high-level, results-orientated work we typically do. So, if we recruit a financial services expert, before launching them into that area, we would typically staff them in a different industry for their first case, like consumer products, in order to broaden their experience. But that means that the specialist/generalist debate is very important to us. We want to encourage people to be what we term 'well-rounded', but we also have to give them sufficiently specialist skills in particular areas to hold their own with clients. As an apprenticeship business, the primary way all consultants develop is by working with those more senior on a variety of clients in a variety of industries. Our consultants typically work on 1–2 projects at a time and change cases every six months. But, in additional to this traditional approach to building one's experience base, at Bain we augment this with three unique features: our system of career management (externships, international rotations, and opportunities with Bain affiliates), our mentoring process, and our global knowledge management systems.

Externships are one way we ensure that our consultants go beyond the traditional notions of 'generalism' and become experts in a wide range of fields. For those people who join us after college as associate consultants, after two years the firm actively encourages them to rotate into a different office, either in the same country or abroad. By that stage, they will have completed several cases (three to six months), usually with a variety of clients, but sometimes with the same one. The usual choice, at this stage in someone's career, is to go to business school, but our aim is to try and replicate both the breadth and depth of learning they would get there. Around 15 per cent of Bain ACs opt for international transfers: I went from San Francisco to Tokyo, to Moscow, back to San Francisco, to Boston, to London and now live in the Netherlands.

Alternatively, an AC may prefer to take an externship with an outside company. Through our network, including alumni and business contacts, Bain will arrange a six month rotation to another company, again to broaden their experience. Under this arrangement, the AC moves off Bain's payroll (and we have very clear rules which restrict a company from hiring them). Both sides win: the company acquires a bright consultant at a comparatively low cost, and our people get a chance to learn about a particular industry and company in enormous depth. In addition, Bain offers unpaid leave of absences to all consulting staff for up to six months if they choose to take a break and pursue other interests such as travel or working for a non-profit organization. For our partners, we have a more formal sabbatical programme which allows them to rejuvenate, take time to write a book, spend time with their family, manage a charity organisation, or even start their own business.

We also have The Bridgespan Group, a separate not-for-profit company which was founded with seed capital from Bain; it is run by a group of former Bain partners who, although they have left the firm, stay connected to Bain & Company by continuing to attend Bain partner meetings. The idea behind the company is to apply the learning Bain has acquired from its work to the not-for-profit sector. It is not *pro bono* – The Bridgespan Group charges its clients fees, but its purpose is to raise the business acumen of the entire sector. The fees charged are lower – something which The Bridgespan Group can do because, when a Bain employee rotates into the group, they agree to take a paycut, but that does not mean we do not have a lot of people who want to rotate into The Bridgespan Group; in fact, it is over-subscribed. People who do rotate in still go on Bain training programmes, and it has the same structure as Bain itself, making the transfer for learning from one organisation to the other much more effective.

Fundamentally, at Bain, it is each employee's responsibility to manage their own professional development: to work with the staffing, to get a variety of case experiences, and to suggest job rotation (either between offices or externships) that will develop their skill base. At the hub of our culture is a belief that one should be 'at cause', not 'at effect'. For example, if you see a problem, it is your responsibility not only to point out the problem, but to suggest a solution. This applies not only when working with clients, but it is also central to ensuring that our consultants are far more 'specialist' than conventional 'generalists'. But that does not mean that we do not have any input into the process: quite the reverse. When a recruit joins the firm, he or she will be assigned a mentor who is responsible for reviewing their performance and discussing career options. People do not usually change mentors (although it does sometimes happen), even when they rotate to an overseas office. Formal reviews take place every six months, but they usually meet with their mentor a couple of times between reviews more informally, just to discuss how things are going, what they might want to do next, and so on. In our view, the mentoring process provides the essential continuity that allows people to rotate through different offices, different countries, even different companies without becoming so dislocated that they – and we – lose any sense of cohesion. It took us years to figure this out, but it does work. It means that everyone has someone to talk to, even when they have just switched cases, moved to another office or changed roles.

The other fundamental role that the firm plays is to ensure that individuals have access to the information and tools they need, as efficiently as possible. In addition to annual formal training programmes, there are 160 modules in the Bain Virtual University (BVU), to which everyone has access. It is another example of how the firm allows its people to pick out what they need when they need it, rather than inflicting a pre-determined learning sequence onto them. We have also developed the Global Experience Exchange (GXE), a knowledge-sharing system which is a testimony to the fact that strategy assignments are now much shorter than they used to be and that clients still expect a high level of innovation. The GXE has five components. Basic tools and techniques are held in the BVU but are fully searchable using the GXE, but the latter also holds: functional information (marketing, supply chain, and so on); industry perspectives; summaries of prior Bain experience; and external data (such as analyst reports). Thus, a single search string can yield all the building blocks needed in a client engagement. It also houses our 'find the expert' search facility, which looks for people rather than documents. This function is called 'peer finder' because it allows people not only to find the 'expert', but to also find a peer who can answer the basic questions.

Within fairly broad parameters, we want people to take control of their professional development, so we provide them with tools and opportunities to broaden their knowledge and experience. They can use modules from the BVU as they need them: they don't have to wait to be told; they can go to the GXE if they want to get an idea of what the firm has done before in a particular area or quiz a colleague. Yes, individuals tend to develop their own areas of interest over time and thereby become specialists in the conventional sense. We think it is important for them to do this because they want to, not because the firm tells them to. That breeds the kind of passion and commitment that is immediately transparent to our clients. **'**

Specialisation: Towards a Value-Based Approach

Is there anything from this that a firm currently caught in the 'generalist' trap can learn? Perhaps the primary message relates to the role of the firm itself. Among niche consultancies, there's little in the way of an organisational structure that gets between the consultants and the markets they serve. They are what they do: management overheads would be irrelevant. At Bain & Company, the firm works hard to give people many options for career development, but leaves the choice of path to the individual. It doesn't decide which people should go where – it lets people make their own choice. Perhaps as far as it can be, it's a self-organising organisation. Market forces apply here just as much as they do among niche consultancies. What the firm does do, however, is make sure that consultants have the information, tools and ability to contact experts which means that, whatever people choose to focus on, they're able to develop a deep understanding of the issues involved very quickly.

The problem for 'generalist' firms is that the firm itself intervenes, allocating resources to particular areas. It plays the role of the market, but can only ever reflect what the market is doing some time after the market itself and it can only ever pick up overall messages, not the nuances and details of how clients are behaving. Much of the effort in HR terms goes into managing this allocation process, not providing it with the tools to manage itself. The key to delivering specialisation without sacrificing the flexibility of the firm does not, therefore, lie in developing a more effective HR function. Quite the reverse, it involves allowing the consultants who work for you to decide what issues their clients face at any given moment and providing them with the wherewithal to solve them.

6

Skills Integration: An Unworkable Model?

[The e-consultancies] attracted a large number of very talented consultants, who were not only excellent in their chosen field of specialisation, but also very well-suited to a collaborative style of working... But these firms also had an impact at a conceptual level. They took a more creative approach to organisational design, and introduced new business models which the industry is still in the process of evaluating and assimilating. Like the development of any new product, there's been an inevitably high failure rate, but that doesn't mean that nothing of value has been left as a result.

Ron Farmer, Senior Director and co-leader of
McKinsey & Company's global Business Building practice

Integrated working was one of the 'discoveries' of the consulting industry in the late 1990s. After decades in which consulting firms – large and small – had divided themselves into functional or industry-related business units, there was a recognition that more value was to be created by enabling people with different skill sets to work more closely together. This value took many forms. Integrated working meant faster working, as tasks which previously had to be done in

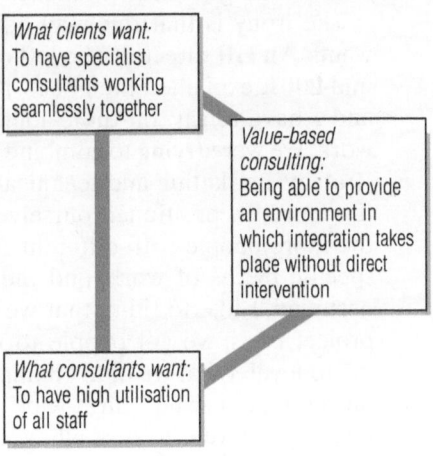

What clients want:
To have specialist consultants working seamlessly together

Value-based consulting:
Being able to provide an environment in which integration takes place without direct intervention

What consultants want:
To have high utilisation of all staff

sequence – the strategy, then the technology – could be done in tandem. It also meant more innovation, as people combined their different perspectives to create new solutions.

'We always ensure that any team – client or internal – has a representative of each of our three core skill sets on it: strategy, marketing and technology', said one e-consulting firm, interviewed in 1999. 'Our "communities" tend to develop strong senses of identity', commented another. 'There's a lot of trust, but at the same time everyone tends to know what's going on. If you're not pulling your weight, everyone will know about it. That doesn't mean that you can take a completely hands-off approach: problems will inevitably arise, but the key thing is knowing when and where to intervene, rather than relying on ongoing intervention.' Integrated working was a pattern that was replicated by almost every substantial new entrant to the consulting industry, and adopted by many of the incumbents as well. As one senior partner in an established firm acknowledged: 'now that the dust is beginning to settle, integrated working is looking like an idea that's here to stay'. That's not surprising, given clients' evident enthusiasm for it. Many people I've talked to have said how refreshing it has been to brainstorm strategic ideas with the people involved in designing the technology to realise them and the marketing campaigns to sell them. 'I'm in no doubt', said one, 'that the strategy we developed was far better – and easier to implement – than if we had gone through the conventional process of working out what we wanted to do, working out how to get our customers to buy it, and then working out how to make it happen'.

The irony is that this may not be a legacy the consulting industry wants. An HR director, who left one of the struggling e-consultancies in mid-2001, explained the issue: 'The trouble is, it's a very tough model, and I have to say, on the evidence to date, I don't think it's going to work. We were trying to combine four skills, the accepted triumvirate of strategy, marketing and technical skills, plus project management. But the more we positioned ourselves as an end-to-end services provider, the more people with different skills we had to recruit, often for very specific pieces of work, and the harder it became to keep everyone occupied. I like to think that we did crack integration on a project by project basis: we got people to work together very effectively at the micro level. What we didn't manage to do was apply the same principle on the macro scale – firm-wide – and utilisation rates fell as a direct result. If we were to start all over again, we wouldn't try to have all the skills internally: we'd form far more relationships with other specialist firms and let them – and the market – sort out utilisation.'

We'll probably never know how much the ideal of integrated working contributed to the demise of the e-consulting firms, but it's clear that it's another example of something that has the potential to create considerable value for clients, but which may also be economically disadvantageous to consulting firms.

Historically, firms have avoiding falling into the trap that the e-consultancies fell into by recruiting more generalist staff as they grew in size. Thus, a consultancy may start off as a niche player, populated by a small number of highly-specialised individuals. As it grows, it expands its coverage of sectors or services, and increasingly looks for individuals who can work across these new areas. As individual markets peak and decline, these people can migrate to more fertile territory, minimising troughs in utilisation as they do so. Thus, the largest consulting firms may not be able to match the 100 per cent utilisation rates of boutique firms that have highly specialised, sought-after skills in emerging markets, but they also don't suffer from the massive slumps in utilisation which the latter typically suffer at the point when the market has moved on. In comparative terms, it's a low-risk, low-return model. By contrast, the e-consulting model was high-risk, high-return – the evidence for both of which we've seen in the last three years.

So is integrated working an unworkable model? Is this something that the consulting industry shouldn't even be promising to deliver?

To answer this, we first need to delve into why integrated working is so difficult to carry out in reality. Part of the reason lies in the different cultures, working practices, aspirations and remuneration expectations of the different groups of individuals involved. 'We had a hard time convincing the strategists that they weren't in charge', said one project manager I spoke to, 'and an equally hard time persuading the technologists to come out from behind their computers and speak to clients directly. And I'm talking about comparatively young people here, not people who've spent half a lifetime working in a rigid organisational model: their attitudes are obviously engrained at an unbelievably early stage – a bit like gender stereotyping, perhaps – and they're certainly just as resistant to change.' Such implicit divisions are echoed by more explicit ones. Our concept of an organisation is founded on structure: we may 're-engineer' it from time to time, but largely what we're doing is replacing one structure with another – the vertical with the horizontal, the centralised with the decentralised. For organisations trying to promote integrated working, these two factors combine to create a Catch-22 situation: intervene too much and you risk establishing explicit organisational divisions; intervene too little and people's prejudices and preferences come to the fore.

In *Evolve! Succeeding in the Digital Culture of Tomorrow*, US academic Rosabeth Moss Kanter identifies six elements which, she believes, contribute to building the organisational 'communities' of the future: a governance structure that balances the need to have someone in charge with allowing individual units greater autonomy; collaboration borne out of shared ways of working; multi-channel, multi-directional communication – face-to-face as well as web-enabled; the presence of people – 'integrators' – who are responsible for sharing knowledge across an organisation fluid and open-ended networks rather than formal committees; strong and extensive networks of relationships at the individual level; and finally, a sense of shared identity and shared fate. 'Communities', she concludes

> are built around people who know each other, understand each other, like each other and have a shared history... Collaboration is not altruism; it stems in part from people identifying with each other and feeling that they share a fate.[1]

The first of these – a balanced governance structure – has not been a problem for consulting firms, either traditionally, or in the last couple of years. While every firm, large and small, has found its own position on what Kanter refers to as the 'continuum from bureaucracy to democracy' – some have taken a slightly more centralist approach, others a more devolved one – almost all have been aware of straying too far in one direction or another. They score highly, too, in terms of their personal networks: as some of the earliest adopters of voice and email, gossip has always travelled quickly in consultancy, although not always in a way that its managers have found easy to deal with. 'Everything in our firm depended on who you knew', recalled one consultant. 'On the day that a massive client win was announced – one of those projects where you'd spend two years in the middle of nowhere, despite it being hyped as a career opportunity – it was preceded by half a dozen voicemails, all effectively saying "don't touch it"'.

But consulting firms have been much less good at the more formalised aspects of horizontal integration – appointing people who facilitate knowledge sharing between business units, doing anything to go beyond traditional top-down communication. Clearly, there will always be exceptions. McKinsey & Company appointed 'knowledge stewards' in the early 1990s to work beside its consultants extracting ideas and information for the firm's knowledge management system.

It then replicated this approach at the end of the decade with @McKinsey, a cross-firm team tasked with ensuring that each area of the firm had access to e-business related skills. The fact that interventions of this nature are comparatively rare in the consulting industry partly accounts for the failure of so many of the e-consultancies to make good on their promise of integrated working. Integration worked well only in small, client-orientated teams where size meant that cross-communications weren't an issue and a sense of common purpose and identity was easy to build and maintain.

One solution may be to ditch the idea of firm-wide integration and limit it to small, focused teams where the conditions are more favourable. To accept, simply, that integration on any more ambitious scale isn't feasible but also to recognise that this doesn't prevent effective integration taking place at a more 'local' level, on a project by project basis. But part of the opportunity for innovation is lost – different skill sets will be combined but only in response to client needs as they arise; doing anything other than reacting becomes impossible. An alternative approach – a third way that some firms have been experimenting with – is where close working partnerships are formed with established players in complementary sectors. This may vary from seconding people in for specified periods of time to using the distinct look and feel of the office space created as a 'hot house' for teams from different companies, bringing them together physically in a way that rarely happened five years ago. The litmus test here has to be the extent to which the firms involved in such partnerships use their proximity to create lasting intellectual capital and/or sustainable value for clients.

But none of these strategies answers the essential question of how you can integrate skills without watching utilisation levels plummet. So, let's look at a couple of examples of companies that are managing to do just that.

Speaking a Common Language at Marakon

'Marakon doesn't come from the same "gene pool" as other consulting firms', says Dominic Dodd, a Managing Partner with the firm. In fact, Marakon was established by ex-investment bankers in 1978. Its founders wanted to take the techniques for understanding valuation in the capital markets and apply them to how business executives could increase their company's value.

‘ We started off concentrating on helping clients to make better informed choices about where to allocate their resources, but it rapidly became clear that there was no point doing this if we couldn't also help clients develop the strategies or choices to choose between in the first place. Also we found that there was no point developing strategies if the organisation wasn't – for whatever reason – going to act on them. As a result, our work has shifted from looking purely at the alignment of resources within a business to also seeing how it can improve its decision-making processes. If a client asks us to look at reducing their warehousing costs, we won't do it; if someone asks us for an entry strategy for the Japanese market, we won't do it. We don't take discrete projects of that sort, where the client already has a set agenda, because, to us, that's really only tweaking shareholder value rather than making fundamental and intregrated choices about managing their companies for value creation.

We are of the opinion that the consulting industry has failed its clients. If you consider the fees that have been paid to consultants over time, can you be really confident that that expenditure has lead to a measurable increase in shareholder value? We'd argue that there are four reasons for thinking perhaps not.

In the first place, an awful lot of consultancy follows a client's predefined agenda –taking their priorities as given or alternatively, coming in with a predefined view about the right agenda: 'you are this kind of client, and therefore you have these sorts of problems and need these types of solutions'. There are very few firms capable and willing to have a dialogue with a client along the lines of 'you're not sure you're doing the right thing, and neither are we, but we'll work with you to try and work out what your agenda should be.' Whether you're a client or a consultant, it's hard not to go in with any preconceptions.

Moreover, consultants rarely challenge clients enough. Of course, you have to make any challenge a constructive one – 'challenging with empathy' is what we call it, but without this, you're never likely to change people's viewpoints.

We also think the consulting industry has become too specialised in how they help clients with their strategies. Consultants tend to specialise in particular business processes or particular industries – after all clients want to be confident that their advisors have specialist knowledge. But with experience these consultants naturally tend to develop clear opinions on what is the 'right' strategy in their field, and they also tend, consciously or otherwise, to replicate this from client to client. That's a sure-fire way to stop anyone creating any value. In that case, what a consultant's really doing is reducing switching costs to everyone following the same strategy in a given industry. A good strategy is a counter-industry strategy, it means doing something your competitors are not doing, and specialisation by industry can be a powerful barrier to achieving this.

And fourthly, there is too much focus in the consulting industry on one-off projects. If you genuinely believe that managing a business is extremely hard, then picking off single issues and solving them discretely doesn't – in our view – add much value. We find that it's more effective to work with a company to improve their decision-making capabilities across the board, so that they're better equipped to make higher quality and faster decisions on an ongoing basis. 'Better facts, not just more facts' is another of Marakon's mantras. Our aim is to develop a deeper understanding of the sources and drivers of our client's value creation than their competitors can, but do so in a way which leads to action and results, not paralysis.

All of this underlies why integrating skills is of fundamental importance to us. We need people who can operate as generalists in different industries to advise clients on a range of issues related to the general management challenge of running a business for value creation. At the same time, we have to avoid the trap of having glib generalists. So we do specialise: we just specialise in the process of managing a company's value. If we have people who only ever know and work in one industry then we'd be in danger of cloning strategy from client to client. We'd be going in with our agenda, not working out what the client's strategy should be. The only way we're going to be able to generate counter-industry strategies is by getting people from different backgrounds, with different perspectives, to work together.

Integration is therefore critical – that's what we're aiming for, and I won't say we always achieve it, but that's the goal we continuously strive to achieve. How do we do it? One way is by having a very strong generic management framework: this is what we train our people in, not consulting skills *per se*. It looks at how and why businesses make money, what kind of strategies they can adopt and how they get their organisations aligned in the drive for better performance than their competitors. This framework has, in a sense, become our *lingua franca*: everyone knows it, everyone uses the same terminology, everyone knows the right kind of questions to ask. It's also fundamental to the way in which we generate choices with clients and help them choose between them – it underpins our entire consulting process. But it also works only because we recruit a specific kind of person: we look for traits – such as being able to 'challenge with empathy' – rather than for experts in a particular field. Taken together, the people we have and our management framework determine our culture as being one in which people work together to deliver integrated advice.

But we also recognise that this is possible because we're still comparatively small. We don't have – and have no intention of having – business units based around industry verticals. Each year, we've set a limit to the rate of growth we want to achieve, irrespective of what the market or our competitors are doing, something we determine quite rigorously based on the ratio of new to experienced staff, and of partners to consultants. We have to be careful that we don't dilute our culture by stretching ourselves too far too fast. There's always a temptation when a new consulting market opens up, to jump straight into it, just as people have been doing with CRM, and did with e-business before that. What you end up doing is hiring people with narrow specialist expertise, whom you can't integrate into the rest of the organisation, so all you've really done is dilute your organisation, not enhance it. As we grow, it's important that people understand what makes us different as a company.

In fact, there's a direct parallel here between what we do internally, and what we do with clients. The type of organisational change management programmes with which we're involved inevitably require the collaboration of different people from different functions. Similarly, the whole idea of 'managing value' involves pulling multiple leadership levers at once, and you can't get people to do that if they're not prepared to co-operate. Integrating skills, therefore, isn't just something we theorise out, it's something that both **9** our clients and ourselves depend on.

Managing Expectations and Egos at Inforte

Like Marakon, Inforte is a small, expanding firm; while its background, services and market are all different, it shares with Marakon the conviction that getting people – consultants and clients – to work across conceptual and organisational boundaries is one of the main ways in which they can add value. The company was set up by several ex-Accenture employees in 1993, initially to focus on the then rapidly emerging client-server consulting market, but foresaw the advent of the Internet.

‘ We set out focused on technology', says Phil Bligh, Inforte's Chairman and Chief Executive, 'but we soon realised that there was an enormous potential market in understanding how technology and strategy interacted. There were – and are – firms that are good at high-end strategy, but what's the point of developing a truly excellent strategy if a client can't implement it? People were developing plans without much understanding of their impact.

We therefore set out to build the kind of organisation that could deliver this. That meant recruiting people from different disciplines who could work together – strategists who'd feel comfortable with technologists, and operations people who'd talk to strategists. Our first lesson was that we had to avoid extremes: the integration we wanted wasn't going to be possible if we recruited the real in-depth technical specialist or high-end consultants. We had to find a middle way. Our next realisation was that client accounts had primarily to be run by people skilled in account management – it didn't matter whether their background was strategic, operational or technical, they had to be a good account manager. And everyone on the team reports to the account manager, not to someone in their own discipline. To stop the account manager being biased – to stop a strategist only thinking about the strategy side of a project or valuing the input of the strategists on the team more highly than the others – we have to invest a lot of time in coaching. It's also important that we don't ask the account managers to pretend to know everything: we ensure that every client team has a full complement of skills, and that there are people there who compensate for the background of the team's account manager. We've evolved the kind of culture that allows people to defer to others with different perspectives when the need arises. So an account manager whose background is in technology won't lose face by taking the advice of the strategists on his or her team. They have to know what they don't know and be able to bring in specialists.

Of course, this means that we need a very particular kind of person. People who are insecure or have too big a personal ego won't thrive in this environment. I'd say that at least half of the success we've had in integrating skills can be attributed to the fact that we've been very careful in who we have recruited. These are essentially people with emotional intelligence. They're not arrogant – clients and colleagues get annoyed by people who claim to be able to do everything. But the other half probably comes from the fact that we also ensure that incentives are designed in such a way as to encourage cross-disciplinary working. The account manager, for example, is responsible for revenue goals in all our skills areas, not just their own.

Integration isn't confined to client situations: we know that, to be embedded in our organisational culture, it has to take place on a broader basis as well. Inforte carries out a lot of research on the interaction of strategy and technology. Michael Porter is a member of our board, and has worked with us to develop ideas about the effect of the Internet on company strategy. Our work with him involved not just those in our IT strategy group, but putting strategists and technologists in the room together. As a result, we have developed a very practical perspective on linking the Internet to strategy, and what emerged was very different from the more theoretical point of view we started out with.

As a result, Inforte is becoming known for the quality of its work, but even more for the quality of its client relationships. Because we're accustomed to an internal model that promotes collaborative working and inclusivity, we've been able to apply the same model in client situations, knocking down some of the cultural barriers between clients and consultants which have traditionally dogged the industry. Account managers are trained and incentivised to make best use of the resources available, and that includes bringing in expert input from the client as well.

I think this has been fundamental in being able to set realistic expectations with clients. One of the recurring problems with consultancy, it seems to me, is that consultants feel under pressure to promise the earth, especially in the preliminary stages of a project – and the disappointment when these promises aren't delivered is almost inevitable. Because we don't have the egos that consultants are commonly renowned for, it makes it easier for us to have a frank discussion about what is and what isn't achievable. You'd think that would make a client more nervous about a project, but, paradoxically, it gives us more scope to find innovative solutions because neither side feels committed to goals it suspects, in practice, won't be achievable. We've therefore built a methodology around expectations management: we put people on client steering committees so that they can sit next to the Chief Operating Officer, for example, quizzing the consulting team about progress to date.

Integrative working, we've realised, doesn't stop or start at our own front door.

Integrated Working: Towards a Value-Based Approach

Skills integration is only possible in an organisation that is held together by a common glue – whether this takes the form of a generic framework, capable of being applied to all clients in all sectors (as is the case at Marakon) or a management ethos that ensures that all parts of an organisation are treated equally (as happens at Inforte). It is also only possible where the market being addressed is clearly defined and circumscribed: Marakon only works on projects aimed at managing shareholder value; Inforte chose the middle ground – the intersection of strategy and technology – rather than trying to develop specialists skills at the extreme ends of this spectrum. Neither company does work outside the parameters it sets. It helps, too, to be working in areas where integrating skills within the client organisation is an important part of the services they provide: the challenges of the consulting firm therefore mirror those faced by their clients.

Perhaps the mistake of the e-consultancies is that they focused on direct intervention (seating people from different disciplines next to each other, for instance) not in indirect intervention (creating the conditions in which integrated working could take place) – and market focus, using integration on the client side to drive internal integration, providing a common way of looking at the world which would outweigh the individual differences between disciplines. And perhaps we're seeing this mistake being replicated by the more established firms trying to move to a more integrated style of working.

As with specialisation (see the preceding chapter), intervention by the corporate entity that is the firm creates inefficiencies: it's because the firm is telling people to work across disciplinary boundaries that overall utilisation rates fall. The solution may turn out to have been part of the problem: direct intervention on the part of a firm makes integration more difficult to achieve.

---

1 Rosabeth Moss Kanter, *Evolve! Succeeding in the Digital Culture of Tomorrow* (Boston, MA: Harvard Business School Press, 2001), pp. 192–3.

7

Innovation:
Getting More for Less

To survive, let alone thrive, in the future, consulting firms are going to have to redefine what business they're in – they're going to have to re-invent themselves. If they're going to provide the highly specialised input their clients are increasingly demanding, and protect their services from being absorbed into the large-scale technology companies, then they're going to have to be focused much more around the process of identifying and exploiting their clients under-used intellectual property.

John Kerr, Managing Partner for Business Consulting, Western Europe, Andersen

'Everybody wants to be innovative at something, but few are willing to bear the risks involved', is how one consultant summed up his experience. Clients vary in the degree to which they want to be leading-edge. Some see it as a matter of pride to think 'outside the box' in all aspects of their business, however mundane. Some focus their innovation on those areas where they believe they have a lead over their competitors – their core competencies – and are content to follow others where less important parts of their business are concerned. No one wants to look ill-informed. But innovation involves risk – risk that you might fail, that the return

What clients want:
To feel confident that they're in the forefront of management thinking (not to look stupid). To be seen as innovators – but in a comparatively low risk way

Value-based consulting:
Creating organisations in which innovation is part of the structure, not an adjunct to it

What consultants want:
To be able to maintain the level of innovation required by clients, but without compromising their profit margins

might be negative, risk that you cannibalise your existing products or lose existing customers.

And a key way in which organisations can manage this risk is to hire consultants. Bring in a consultant and you – in theory – get access to leading-edge thinking on a particular subject, but from an organisation or individual that has either already researched this field and has therefore accumulated some knowledge about what will and won't work, or has an established process and/or skill set which minimise the risks of researching a completely new topic. In other words, prior to winning the contract, the consulting firm has already had to make an investment – in researching a particular field for itself or in building a core-competence to research topics on behalf of its clients. Either way, it's the consulting firm that's taken the lion share of the risk. An organisation can further reduce the risk it faces by drawing up a contractual relationship with the consulting firm (payment by results, and so on) that pushes responsibility for achieving success in that specific instance back to the consulting firm. Of course, it's dangerous to generalise about all clients: there are undoubtedly some who are both innovative and willing to take risks; equally some who are both risk averse and have no aspirations to be innovative. But they are both in the minority: the bulk of clients – at least according to my research – want to have their cake and eat it, to have some degree of innovation (like Goldilocks – not too hot, not too cold) but not bear the risk (Figure 7.1).

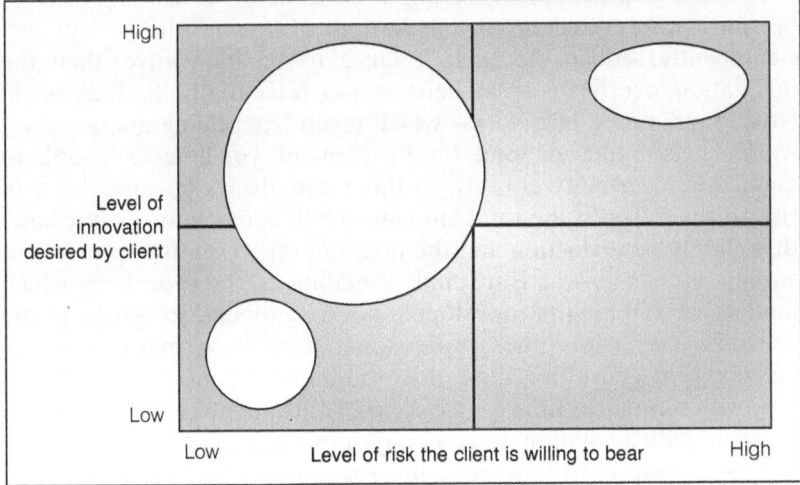

Figure 7.1 *Distribution of clients according to their innovation/risk profile*

The consultant's view is identical – from their own perspective. A consulting firm wants to supply clients with the level of innovation they require, but use consulting projects to develop new ideas and, indeed, hone the firm's ability to develop new ideas. Just as clients bring consultants in to reduce the risk, so consultants want client projects because the fees earned fund part of their research and development. In an ideal world, clients would recognise that doing something new requires more input from the consulting firms: innovative projects would cost more than routine ones – the consulting firm would acquire innovative thinking while bearing less of the risk.

An important part of the client-consultant relationship is the way in which risk is balanced between each side. This applies both to financial risk (that there will be no return on the money invested) and delivery risk (that the project cannot be successfully completed): each side wants to offload as much of both these risks to the other side as possible. And, because it's a trade-off, the results are rarely completely satisfactory to either side. If the consulting firm ends up shouldering most of the risk, it may look to cut corners – be less innovative – in order to reduce its exposure. If the client bears most of the risk, it may well breed resentment at the role of the consulting firm (that criticism of consulting – that consultants take your watch to tell you the time then invoice you for it – must spring from this situation).

It's an intractable problem – or is it?

The first step towards resolving it has to be for consulting firms to become more efficient at innovation. If a consulting firm can significantly reduce the time it takes to be innovative, then the innovation 'overhead' on a client project (essentially the increase in costs – and, potentially, fees – which result from doing non-standard work) is also reduced. Some tension remains: you'll never be able to eradicate uncertainty entirely, so there will always be some level of risk to share. But if the total amount of risk is much lower, it follows that client dissatisfaction and the pressure on consulting firms to cut corners are also lower. But simply speeding up the process by which innovative thinking is developed doesn't appear to work, partly because – as with other professional services – innovation and efficiency in consulting have not traditionally gone hand in hand. Innovation requires time – time to research, analyse, reflect – time to come up with a solution. Indeed, it requires unlimited time – because you don't stop until you come up with an innovative solution. The argument goes that putting a deadline on the activity, or trying to perform it more quickly, necessarily constrains innovation. The consultant is like an artist. The analogy certainly applies within the advertising industry. 'We're falling between two stools', said one advertising executive I spoke to. 'Our core competence – our heritage – is in the creative process of designing ads which sell. But, as the industry becomes more competitive and clients more demanding, we're finding ourselves under constant pressure to do what we've always done, only more quickly and more efficiently. As a result, we've acquired processes and performance measures that don't sit comfortably with our traditional way of working. But we don't have any choice. Does that make us less creative? I think it has to.'

There are two other problems in concentrating on expediting the innovation process. First, clients tend to equate process efficiencies with reductions in fees. During the 1998–2000 e-business boom, many consultancies found themselves having to reduce the length of time it took them to deliver projects because speed to market had become clients' number one priority. Thus, strategy assignments that, conventionally, might have taken several months to complete were shrunk to as many weeks. This was largely achieved by breaking projects down into more distinct elements, each of which could be worked on simultaneously, rather than sequentially. Consulting firms claimed that it took just as many resources as the traditional way of working and wanted to charge the same price; clients saw it as a reduction in the amount of time committed by a firm and wanted a commensurate discount. The second problem is that it causes increasing internal divisions within the consulting firm itself. A large proportion of efficiency relates to familiarity: we can all do things quickly that we do often, whether that's commuting to work or making a meal. The more efficient we want to be, the more familiar the task has to become – and the less easy it is for anyone to do. Consulting firms face a dilemma: by streamlining their innovation processes, they will end up restricting who can undertake these processes to a select few, and these people will, in turn, become distanced from the mainstream organisation. The division between product development and product delivery has always been a difficult one for consulting firms. Countless products have been developed but undersold because those involved in sales and delivery weren't involved in development – the 'not invented here' syndrome. But the opposite approach – ensuring that product development is fully integrated in the consulting business units – is highly inefficient: people are constantly pulled off onto projects, their training is focused on delivering rather than development, and so on. Creating too separate a product development process, and too distinct a group of 'product developers', may mean that you'll end up with more projects but you'll find it harder to motivate the rest of the organisation to sell and deliver them. Ironically, getting better a product development may reduce your sales.

So you can't improve the efficiency of innovation simply by improving the innovation process – although this is effectively the strategy that many firms have adopted since the mid-1990s. You need to do something else. But what?

A few years ago, the analogy with artistic creativity would have applied equally well to tax consulting. Tax consultants typically did one of two things: complex but essentially standardised tax calculations and the interpretation of tax rules. While the first of these activities is well on the way to being replaced by software packages, the second has remained the preserve of high-charging tax consultants who have needed to apply considerable creativity to developing options and ideas. At the same time, tax advisory firms have been faced with the growing problem of how to make better use of these expensive resources. They have needed to speed up the creative process so that it can be applied to more clients in a shorter space of time, and technology has played an important part in enabling them to do this. The software packages now coming on the market provide much more sophisticated modelling tools: they don't replace the creative process entirely, but they do allow part of that process – testing and evaluating a wide variety of complex scenarios – to take place much more quickly.

But although technology appears to provide a solution to the problem of sustainable innovation, in reality it's simply fixing one of the symptoms, not the root cause. Technology effectively provides an opportunity to make creativity tools more accessible – it democratises innovation by deskilling the process. You don't have to be a strategy whiz to have good idea, you can use an creative technique. It's less the process of innovation that's inefficient than the fact that it tends to be carried out by only a subset of people in any organisation as a whole: the costs come in having to mobilise an organisation to adopt an innovative idea that they had no part in creating. The key to sustainable innovation, therefore, has to lie in building an organisation where everyone believes it's their job to be innovative.

Different by Name: Different by Nature?

Differentis is a European consulting firm which is facing exactly this challenge. Founded in early 2000, at the height of the e-business boom, it's survived where others have failed due to a combination of secure funding and professional management expertise (largely from CSC Europe). It sees being able to generate innovative ideas that can be delivered efficiently and effectively as one of its fundamental purposes.

Ron Mackintosh: What's different about Differentis? There were two reasons why we set the company up. The first is my belief that the Internet is an extraordinarily disruptive technology, whether it's the Internet *per se* or just the connectivity that it creates. This new era of connectivity changes a lot of the fundamental rules about how businesses operate, and the way they work. It creates, I believe, an enormous opportunity to transform the way that current business is done. Differentis therefore focuses on transforming business.

The second reason is quite a different one. It's the belief that there's an opportunity to create a powerful European player in the consulting marketplace whose approach is driven by a very strong European philosophy. The American way of consulting, although effective, doesn't have to be the only way. I often ask myself why are American firms so dominant in the field of consulting? I don't think there's any specific reason, except people have just allowed it to happen. While recognising the strengths of the American model I would like to see a different model, a European model. American consulting firms have tended to be much more enchanted by the new, by the innovative, and there's always been a one- to two-year lag in those ideas being disseminated in Europe – and that's something we'd like to change.

Bruce Rogow: We're therefore trying to create an environment where we aren't under pressure to recommend a given solution to clients: we don't have any vested interests and we don't have a massive organisation we have to support. There's an awful lot of arrogance in the consulting industry – consultants who come back to the office and openly mock their client's ignorance. But I remember a time when people used to be excited that consultants were coming: there was a real belief that we were going to bring fresh ways of thinking and stimulate discussion. Consulting ought to be a novel experience, something that people remember. In the past, some firms have been very good at doing this: it wasn't just that they had interesting things to say, but they found ways of presenting that information that made clients sit up and pay attention. Consulting was an experience.

Ron Mackintosh: But, at the same time, we didn't want our innovation to become prohibitively expensive. I don't think clients get a great deal from the current consulting integration businesses: there are some scale benefits to the mergers that have taken place over the last few years, but I think they're outweighed by problems that scale brings. We've come to a point where we've built bigger and bigger consulting firms, where we've gone beyond the value we're actually bringing to clients, so much so that clients are now paying premium – not low – prices to the biggest firms. In almost every industry scale brings economy – except, apparently, in the consulting industry. Finding clients who are totally satisfied with their existing consultants, and the value they derive from that relationship, is a rare and very real issue in today's marketplace.

Bruce Rogow: There are several different kinds of firms as I see it, each of which has a different type of innovation. As the name suggests, 'market makers' have primarily been innovators, simply put – they're designed to make markets, often for quite finite periods of time. But one of the side-effects of market-making is that you create hysteria, a hype – something that, with e-business, has taken on a tone of lunacy and a complete lack of professional integrity. What you've got in the market-making segment is a bunch of Hollywood-style promoters: it's become the Don King school of management consultants. Everybody's trying to be a market-maker; everyone's after the next big new thing. And I think one of the problems is they've gone too far – they've effectively trivialised market-making.

'Trade winds' consultants tend to go with what the current issue is and recapitalise on that – Y2K was a good example, but there's a third category of firm that comes up with terrific ideas with which they fall in love. Although initially innovative, the level of innovation in these firms shrinks, the more they become attached to their original idea; ultimately, it disappears completely. There's no deliberate attempt to stifle innovation here, but there is a perceived need to standardise – the process people take over. The proposition of 'Home Depot' consultancies is: 'you can go ahead and hire someone from one of the traditional firms for half a million dollars to study something, or you can buy a research study and talk to our analyst almost any time you want for $20,000'. You're probably going to get a better answer from the consultant, but for just $20,000 …

Each of these types of firm is innovative in its own way. For the Home Depot consultancies, for instance, innovation lies in the way they commoditise their services and deliver them at negligible costs. Other firms innovate around their 'socialisation': they're built on the relationships that they've established – and this is where they have to be innovative, not in their products or services. For other firms, it's innovation around process that matters most, something that requires considerable organisational experience – you need to understand how management learn and you need to be phenomenally consistent in the way you apply your lessons. That's a form of innovation.

I see Differentis as a firm trying to make a market. It started off with an understanding of where the market was and where it might go. It not only understood today's context but had a good grasp of what tomorrow's context would look like and what drivers would create it. By positioning Differentis ahead of the curve we are in a fortunate position: when a client asks who our competition is, we can say we have none.

Nick Shelness: We've reached the point where we've developed a core concept – the joined-up business. Up until now, the primary globally integrated business model has essentially been an imperial one: globally integrated companies have been global empires in which everything has to be done in the same centrally dictated way everywhere. This model has been reinforced by information technology vendors whose business models required them to lock their customers into a single technology base – theirs has been an anti-joined-up strategy – and for their own very good reasons. What's interesting about the environment in which we now find ourselves is that IT vendors will no longer be able to make money by doing that, because to a great extent they've saturated that market – in some sense the need that existed has been met. Moreover, having a system that serves the needs of one corporation but can't serve the needs of others isn't going to be a sustainable strategy in the future. Businesses will need open systems that can talk to multiple organisations and that's forcing a wholesale change in the way that vendors think about their products.

This shift is, in turn, making a much more heterogeneous business model – the joined-up business – a real possibility. People and organisations will be able to do things differently, while still being linked to a collective entity: the corporate and the individual won't be incompatible. What's, therefore, so different about joined-up business is that it represents a new socialisation of work, a way of networking, of bringing people together that promises to be extremely powerful. Organisations will be able to expand without having to ensure complete standardisation; people will join networks because they choose to, not because they have to.

Bruce Rogow: Another issue is to ensure that we understand how technology fits into the blueprint we're building of the joined-up business. You have to be – and clients are – very sceptical about the role of technology. I was born in the atomic age: everything was going to be nuclear – watches, telephones, cars. But it didn't happen. Right now we're at a very critical point because there's a family of technologies out there with no clear leader. But, unless we put something really specific behind your ideas, in terms of what the applications are or how they're going to be integrated, then it's difficult for people to envisage what you're talking about. Technology is valueless unless it's applied.

Paul Seaton: We're now in the process of converting these ideas into concrete propositions we can market effectively. When you start to unpick what the idea of the 'joined-up' organisation means in practice, you find a lot of issues around technology integration, both behind the scenes in terms of making disparate legacy systems talk to each other, but – crucially – also at the front end, where you've got to develop interfaces that take into account the fact that different people do things differently. Thus one critical issue for us is around 'user experience', by which we mean finding appropriate ways to integrate technology into people's ways of working. By combining business and technological robustness and a consideration for the user experience we have created a real diffentiator for us and a real offering for the market.

Ron Mackintosh: Part of the key to sustaining our level of innovation is to find the most interesting clients – people that are willing to allow us to share their problems with them. We have to respect them for what they've built and what they've created in their companies: the only reason we exist is to serve these people. But remaining focused means that we've got to have the guts to say no when they ask us to do a piece of work which we don't believe is the right thing to do.

Nick Shelness: Consultants haven't learned a lesson that software vendors learned years ago. When you go into a potential client, you acquire enormous credibility if you sometimes say, no, our product won't do that. Claiming your product does everything only breeds distrust. So saying 'no' can be a very powerful selling technique, but most consultants are frightened to employ it.

Ron Mackintosh: Strategy is all about making a choice, a choice about what you will and won't do. In some ways, the act of choosing can be more important than what you choose to do. We see a lot of people investing huge amounts of money in the joined-up business idea, and at the technology level there's a whole set of software companies now creating the tools needed to realise it as interconnectivity software gains momentum. So we think the timing's right.

Nick Shelness: The next challenge for Differentis is to build an organisation around this idea. That's going to take a lot of discipline: you have to be very precise, and that's harder than coming up with the idea in the first place. I've seen companies with highly informed, innovative leaders possessed of deep insight, leading a comparatively ill informed (effectively illiterate) organisation. What you get at the lower levels is homogenisation, people not thinking for themselves, being jacks of all trades but masters of none. One of the most important things Differentis has to avoid is falling into this trap. If you've got innovative ideas, you have to disseminate them throughout the entire organisation. And you have to find people who are sufficiently specialised to appeal to clients, but broadly-based enough to be able to communicate with each other effectively. We can't send in technologists and business specialists who can't talk to each other: plenty of people are already doing that!

That, in my mind, is the dilemma of what Differentis needs to try and do: to have deep insights and deep skills, but, at the same time, an ability to mesh it all together so that you end up with a totality that is greater than the sum of the parts. That's something that would be unthinkable in a larger organisation.

Bruce Rogow: A friend of mine used to say that organisations die of terminal 'fungis'. I didn't know what fungis was so I looked it up in the dictionary, but there's no such word. So I asked him: 'what does it mean?' 'It's a management cancer', he said, 'people become fungible. Anyone can do anything. Take such-and-such an executive: he's a good guy – it doesn't matter what division he's in. You're a telecommunications expert, so we're going to put you in charge of the software packages.' That's terminal fungis, and it stifles creativity.

Ron Mackintosh: It's essential that we're consistent about our message. We have to make sure that we're delivering it in a way that people understand and can apply. We have to be highly disciplined and rigorous, and we have got to stick with it. The problem that often comes with very bright people is that, once they've got a great new idea, they just want to go off and find the next great new idea. What we want to do is help our people really understand what the joined-up business idea means and how you make it happen. We've got to keep very, very focused on this – it's the heart of what we're about.

Nick Shelness: Ron has the very difficult job of finding a balance between people who want to be innovative and can't finish a sentence or an idea, and those who know how to run a project. It's like herding cats. If you have ever observed cats, you'll know that you can't herd them. You have to get them to do things for different reasons – you have to create something that looks like mice. The consultant's equivalent of mice is clients: management in the consulting business requires a skill of being able to create the conditions where bright people can come together and share their individual perspectives – and that's what happens when you focus them on solving clients' business needs.

But, if we're successful in winning the most interesting clients, we have to match this with recruiting interested people, people who just love to understand how a company works. When you start to understand that, you realise how complex things are, how much skill has gone into pulling it together, and that means that you've got an opportunity to see where you can really help them to take the business forward. But there's an overhead there from a client point of view: you're not coming in with a smart, immediate solution to what-ever they think their problem is. Instead, you're challenging them to take part in a process that may be more difficult – and is that really what they're willing to buy?

Bruce Rogow: But curious, creative people are also difficult to manage because what you have is, in effect, a portfolio of different skills, rather than a conventional, homogenous organisation. Another challenge is to ensure that all the people in a heterogenous organisation like Differentis understand how they contribute to common goals. And one of the problems the consultants have had is that often they don't feel they make a difference, either within their firm or to their clients: the legacy they leave doesn't last.

Paul Seaton: What we've tried to do is involve as many people as possible, if not in the research and thinking about the joined-up business idea, then at least in the way we translate it into service lines – which are, after all, the core of our business. We're developing what we call solution sets – a process where we map aspirational points of view against our capabilities – and this creates a flow through from the generation of an idea to delivery. It's internal incubation, and that creates a 'pull' for new ideas as well as a mechanism by which ideas can be 'pushed' out to the rest of the organisation. That being said, it's impossible to escape a degree of tension between the people who're responsible for thinking through the implications of concepts like the joined-up business at an intellectual level, and those who are tasked with turning these ideas into commercially sustainable products. We're still learning how we balance this, although the process we've developed seems to be working well so far, and one of the lessons, I'm certain, is that creating an organisation that not only can innovate, but can continue to innovate, requires people who've got their own, strongly held opinions at all levels.

Innovation: Towards a Value-Based Approach

If there's one thing that singles Differentis out from most other consulting firms, it's that the entire organisation is being built around the innovation process. Innovation isn't just something that is being left to expert outsiders thinking great thoughts: their ideas are being challenged and developed by people at all levels within the organisation. For most consulting firms – strategic consultants are the exception that proves the rule – innovation tends to be something you go off and do: it's not what consultants do in their everyday lives. Innovation is inefficient, only because it's perceived as a separate process, an overhead, if you like, only tangentially related to the core consulting process. Becoming efficient at creativity involves, not innovating less (what consulting firms often want) or faster (what clients always want): it involves embedding innovation in everything you do.

8

Data, Information and Knowledge: Re-Engineering the Intellectual Value Chain

The big threat we see is distintermediation ... If you put together several trends – the greater availability of information, the rise of the 'free agent' worker, the role of talent agents – the traditional consulting model could find itself under a lot of pressure.

Christoper Meyer, Director, Centre for Business Innovation, Cap Gemini Ernst & Young

In *Profit Patterns: 30 Ways to Anticipate and Profit from Strategic Forces Re-Shaping Your Business*, a team of consultants from Mercer Management Consulting argue that

Industry value chains used to be incredibly stable. Today, those value chains have been compressed, broken up and put together again... Within this context, companies are using new, non-traditional criteria for value chain moves – moves driven by improving return on capital employed, achieving strategic control, creating business design innovations, and capturing the customer relationship.[1]

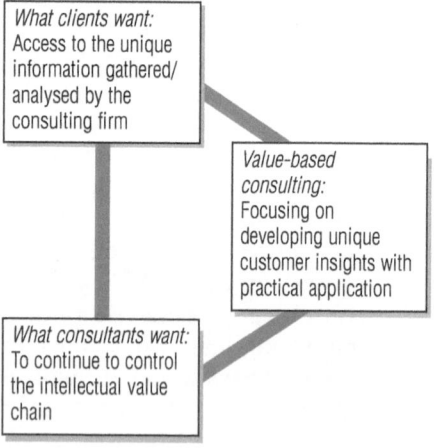

What clients want:
Access to the unique information gathered/ analysed by the consulting firm

Value-based consulting:
Focusing on developing unique customer insights with practical application

What consultants want:
To continue to control the intellectual value chain

Among the strategies they highlight are: 'deintegration' – the shift from vertical integrated business models to value chain specialists, companies that chose to become highly competent at just one part of the value chain rather than the value chain as a whole; and 'reintegration' – where companies expand the scope of their business model up and down the value chain in the search for shifting profits and power.

Intellectual capital – its massively increased proliferation and availability – is one of the primary forces that's threatening to de-stabilise the 'intellectual' value chain in the consulting industry. But is this an opportunity for consulting firms to reposition themselves, offering new services to clients – to reintegrate. Or is deintegration a threat? What can consulting firms learn from their own analysis?

Clearly, the answer to this question varies, depending on your position in the market. Some consulting firms specialise in explicit, structured information – the more operationally-based firms; others in implicit, unstructured knowledge – the more strategically-orientated firms. This chapter will examine how changes in the intellectual value chain of consultancy may impact both of these groups.

Consultants as Infomediaries?

Information is the world's most recent dependency, and it can feed its habit 24 hours a day, seven days a week, at work, at home, on the move. 'It's scary', I remember a consultant telling me in the late 1990s, just as awareness of the Internet as a source of information had gone through the roof. 'As a consultant, you like to think you're going into a client's office well-informed, certainly better informed than they are. Why else would they want to talk to you? But we're suddenly finding that the client's picked up a bit of news ten minutes before and is looking to us for an instant reaction to something we know nothing about.' But client empowerment rapidly gave way to information overload. Just six months later, clients were complaining that there were now so many sources of information available that staying ahead of the game had become a full-time job in its own right. 'With so much information around', said one, 'we're really looking to consultants to filter out what's important and send us information about new ideas that might be of interest to us'.

In *Net Worth: Shaping Markets When Customers Make the Rules*, John Hagel and Marc Singer mapped out what such a role could look like:

> In order for customers to strike the best bargain with vendors, they'll need a trusted third party – a kind of personal agent, information intermediary, or *infomediary* – to aggregate their information with that of other consumers and to use combined market power to negotiate with vendors on their behalf ...The infomediary's role will, in fact, be a very traditional one

... They'll create a new form of information supply. By connecting information supply with information demand, and by helping both parties determine the value of that information, infomediaries will build a new kind of information supply chain.[2]

Although Hegel and Singer were primarily discussing inter-mediaries between consumers and suppliers, their comments about the market conditions most likely to give rise to infomediation – uncertainty, fragmentation, multiple sources of information – are all ones that would apply equally well to the market for management ideas that ultimately drives consultancy. The interesting question is why infomediaries haven't appeared.

To understand why, we have, I think, to look at the nature and evolution of the intellectual supply chain as it applies in the consulting industry. Almost every type of consultancy – from the most strategic to the most operational – has its own hierarchy of raw data, information/analysis, and applied knowledge. A strategy project may involve collating data on a particular market, analysing pertinent trends and formulating a plan of action based on the interpretation of the results in the light of previous experience. A project to upgrade a company's computer security system might start with collecting data about the individual security of its different applications, known infringements, and so on, move on to a gap analysis of what there is against what is required, and culminate with the purchase and implementation of the appropriate package by specialist techno-logists. Over the course of the twentieth century, the focus of consultancy has gradually moved up these hierarchies (Figure 8.1). The consultancies, like McKinsey & Co and Booz Allen & Hamilton, which emerged out of the scientific management movement, initially concentrated on gathering data through time and motion studies. By the 1950s and 1960s, as data was increasingly captured by computers rather than people, consultancy had moved on to information analysis. When this too was superseded – by spreadsheets – the industry moved on once again to the application of that analysis in a practical, human context – knowledge. Indeed, by the 1990s, endemic client dissatisfaction with consultants' lack of accountability was pushing at least some firms beyond knowledge into implementation, thus effectively taking consultants out of the world of theory and into the real world of work.

For a long time, consultants succeeded in maintaining their grip on the data and information components of the overall intellectual

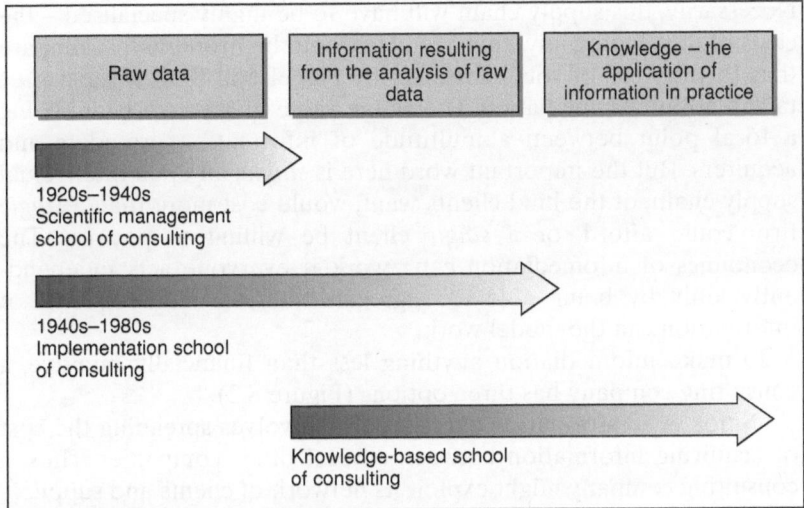

Figure 8.1 *The changing involvement of consultants in the intellectual value chain*

supply chain: they had the resources to seek out data; the frameworks for effective analysis; the track record which provided a basis for interpretation. But, by the time I was talking to the consultant quoted above (1999), this grip had been lost, new companies (specialist research companies, for instance) had emerged to challenge the consultants' role in this respect; search engines allowed clients to trawl through the reams of data they had previously paid consultants to do. Some firms hit back, seeking to re-establish their authority by creating unique sources of data – consumer panels, benchmarking databases, and so on – to which, they could be sure, clients would not have access.

The problem with this strategy is internal: can you do this profitably? Clients want an 'open' system. They want the world to be like it was, when they could rely on the good consultants to be well-informed about a wide range of business topics, even if that 'well-informedness' now has to be applied to a vastly wider range of topics than was true in the past. From the consultants' standpoint, this just isn't possible, or at least not without pouring in more resources than a single firm has access to or that a single client is willing to pay for. What consultants want is a closed – indeed, hermetically sealed – supply chain to which only they have access: only this can guarantee that the intellectual capital they take to their clients is unique.

Necessarily, this supply chain will have to be highly specialised – the costs of developing anything broader would be prohibitive. Compare that to the kind of role envisaged by Hagel and Singer when they talked about infomediation. The whole value of this role lay in being a focal point between a multitude of information providers and acquirers. But the important word here is *single*: an open intellectual supply chain, of the kind clients want, would cost more than a *single* firm could afford, or a *single* client be willing to pay for. The economics of infomediation can't work if everyone acts independently: only by being able to aggregate sources and recipients of information can the model work.

To make infomediation anything less than financially punitive, a consulting company has three options (Figure 8.2).

Option A – information aggregation – involves spreading the cost of acquiring information between several other companies. Thus, a consulting company might exploit its network of clients and suppliers to pull together information in highly specialised areas. A consulting firm focusing on the telecommunications market could therefore pull together, ideally on an automated basis, selected information from telecoms companies, customer surveys, regulatory authorities on behalf of a particular client – the Intranet equivalent of the extended organisation. Option B – client aggregation – is the converse of this: the consulting company itself would invest in building up a proprietary source of knowledge which it could then either sell (for a comparatively small amount of money) to a large number of clients, or use as a marketing tool with these same companies, opening doors that might otherwise have been closed. Option C – infomediation – combines Options A and B. The consulting company would collect re-usable intellectual capital among its customers and suppliers, supplement this by developing its own information sources, and then sell or market this to other companies.

All three of these options are possible – though Option C is perhaps the most sustainable, although also by far the hardest to achieve.

The 'Downstream' Value Chain

But what of the opposite approach? If moving 'upstream' in the intellectual value chain (back into information and raw data) is problematic because it's a strategy that hard to sustain, what about

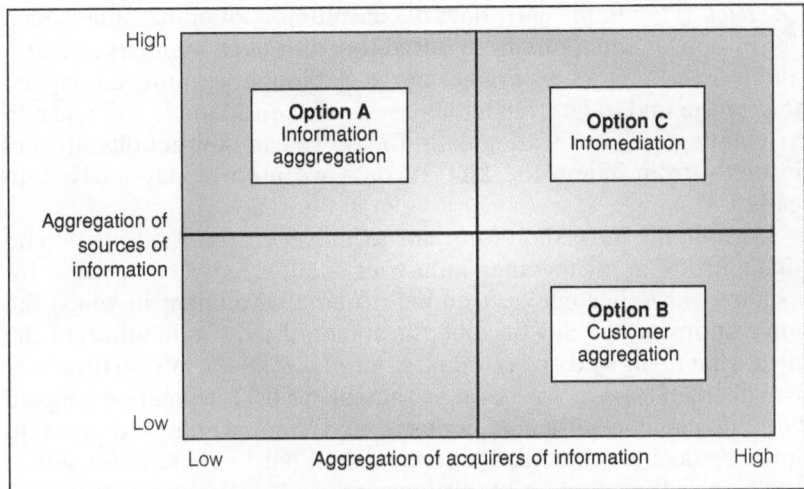

Figure 8.2 *The three options for infomediation*

moving further 'downstream' (into knowledge)? This is, after all, what the consulting industry has done historically. Does it really matter that the grip it had on the data and information components of the value chain, which was in any case already weakened, now disappears completely? There's certainly a danger that, without some level of intellectual differentiation between clients/suppliers and themselves, some consulting firms have found themselves skating on thin ice: commenting on issues of which they only had a high level of understanding; implementing systems they were not expert in. As one client put it: 'we've had a stream of consultants coming to us, all claiming they've worked on the same turnkey project in our industry – which just isn't possible – almost all of whom, we suspect, are dressing some comparatively limited know-how in fancy terminology and glossy brochures. What we find really amazing is that they think it's a credible approach.' Or another: 'why should we hire a firm to implement a particular package when the software development company have their own consulting division? We've already taken the decision to buy the package, so objectivity isn't an issue for us. But expertise is. To us, it's just not credible that a broadly-based consulting firm can offer the same depth of expertise that the original development company can.'

So how, then, can a consulting firm secure its 'upstream' position? How does Mercer Management Consulting apply its thinking on changes in the value chain to itself?

' *Rick Wise:* In the early days of consulting, a lot of the value added
by consultants came from our ability to crunch numbers in a grid
and to hand-plot an experience curve. Although we'd moved beyond
this by the mid to late 1980s, there was still considerable value just in
assembling and aggregating data. Today, we can now get directly and
instantly from Yahoo the kind of data we used to pay analysts to
gather.

Consultants have therefore had to move up the food chain: and
that's no different to other industries – automotive companies, for
example, have become accustomed to an environment in which the
innovations of one day become the standardised commodities of the
next. That being said, the raw data itself, and even the information that
was distilled from it, was never as valuable as the information coupled
with practical experience, contacts and frameworks and used to
improve decision-making. This has made it hard – if not impossible –
for a consulting firm to be disintermediated. If you were only ever
packaging information to begin with, then you were already on thin
ice.

But that's not to say that the consulting industry isn't facing some
significant challenges, as a result of more information being freely
available. One issue we're now seeing is a compression of the time to
do what we're being paid to do. Clients are now driving us towards
very tight deadlines: you can no longer charge as much for this work
as you used to. That's a problem, but it's not the same problem as
Yahoo cannibalising the role of the consultant. Most of the attempts
to act as a third party infomediary have not come to much so far: the
real threat is of changing economics, not of disintermediation. Like
the advertising industry, we're certainly in an environment in which
clients know more, but this isn't so much because of their access to
new sources of data, but because we've spent the last twenty years
training many of their staff – they're *alumni*.

Richard Balaban: At Mercer Management Consulting, our job is to
help create competitive advantage that's actually worth something in
the marketplace. There are three characteristics of genuine
competitive advantage. First, it must mean that you can do a better job
of servicing your customers' needs – that applies whether you're a
paper manufacturer or someone who sells floor space to retailers, it
cuts across all sectors. You have to be able to identify what your
customers actually need and encourage them to pay more to receive
it. Your customers need to be able to say you do things better – and

it's important that it's your customers who do this, not an analyst or research company: the specific benefits of what you offer have to be both material and measurable from the customer's point of view. Second, your competitive advantage has to include a management capability – to understand what goals you should set and how they can be achieved; to build an organisational architecture that gives confidence you can deliver on your promises. Finally – and, in a sense, this is a product of the first two characteristics – your competitive advantage has to be recognisable in the capital markets, through a combination of financial performance that is better than your peers, plus the expectation that that superior financial performance will continue in the future. Frequently companies that are good at one thing lose sight of the fact that they have to evolve in response to changing market conditions: they don't see the need to 'migrate' their value, and that's when investor confidence starts to fall and when their shares get discounted.

Very few companies exhibit all three of these characteristics: there are perhaps just one or two of them in any industry at any given time. And, in some industries, there are none. In chemicals, for instance, no company has ever managed to break away. Why? Because they all share the same information. Information is readily available – it's a commodity. This creates an industry 'disease' in which standard patterns of behaviour emerge, and where the proposition made to customers is essentially the same across the board. You see it in the automotive industry, for example, and in many manufacturing sectors; we saw it repeatedly during the dot.com boom, where everyone was offering the same thing.

We can see this too in the advertising industry. In its heyday in the 1950s, you had the big names sitting on top of mountains, creating – really for the first time – images of things that people loved. Then, in the 1970s and 80s, you had the continuing wrangle of effectiveness. But, latterly, something's shifted again. Skills that used to be the sole preserve of the agencies have migrated to their clients. Similarly, information which used to be gathered by the agency, is now being gathered by the client. The ability of advertising companies to do anything more than the creative part of the process has virtually ended – and we've seen industry-wide consolidation as a result.

John-Paul Pape: There are parallels here with the consulting industry. Part of the role of consultants has been to create and disseminate information, and we've reached the point where there's too much

information, all of which is equally accessible to all the participants in
the consulting value chain, from our suppliers to our clients. Look at
how many companies are sitting around feeling nauseous about the
millions of dollars they've spent on implementing large scale systems
without reaping anything like the expected return. They've all been
trying to do what they know their competitors have already done.

Richard Balaban: It's what I'd call 'keep-up' consulting. It's where
consulting firms, relying on the same sources of information, have
become too similar: they've got nothing original to say to clients, and
that translates into developing 'keep-up' strategies for their clients.
There's an enormous difference between that and the kind of
consulting that provides a unique insight into a client's competitive
position. And I think it's important that we don't underestimate what
an enormous dilemma this is for managers who are responsible for
increasingly complicated value changes and arrays of very different
customers. It doesn't matter if you're a clothing manufacturer or a
multi-sector industrial company. This is why strategy consulting has
become more important again – because we offer a distinctive point
of view that more operationally-based consultancies, increasingly
commoditised, can't.

You could divide the sources of information in the consulting
market into four categories. There's readily available information –
the kind of thing everyone knows. This varies from industry to
industry: there are some sectors where information is comparatively
scarce, and others, like the airline and retail sectors, and typically
asset-intensive sectors, where everybody knows just about everything.
There are things that a company knows about itself and its own
operations, and the things it quite legitimately knows about its
competitors. Finally, there are the kinds of insights into an
organisation's particular circumstances that only ever come from
talking to customers. At Mercer, we start with getting information and
insights about our clients' customers, and their customers' customers
(and so on), and analyse the extent to which their proposition offers a
genuine competitive advantage.

Rick Wise: We're certainly seeing consulting firms shifting towards
offering proprietary data and unique interpretation – largely
information gathered from their clients' customers and analysed using
specially-developed tools. In fact, some types of data gathering, which
we used to find prohibitively expensive, are now much more
affordable. Mercer does a lot of online surveys, for example: they're

cheap to set up, information can be gathered in a week, and we can even look at interim results every hour or so. That's quite different to the conventional $300,000 survey which took three weeks to organise, three weeks to execute and three weeks to analyse. In terms of external data sources, there doesn't seem to be the consulting equivalent of a Bloomberg – that kind of very valuable, proprietary external database. There are some firms that have specialised data, but you have to remember that the consulting industry is a collection of fragmented areas of knowledge. It's possible that single firms could dominate different segments of the market, but it's hard to envisage one being able to dominate all of them in the way that – say – a Neilsen can in the advertising world.

John-Paul Pape: If there aren't going to be unique sources of information, we have to find unique approaches that can be applied to the data, if we want to differentiate ourselves. You only have to look at what's been happening in financial services to see the importance of this. Most financial information – financial accounts, SEC filings, and so on – now comes from one source: Thompson. There used to be a myriad of companies in this field, but there's been a massive amount of consolidation as the price customers were willing to pay for this data fell. Companies had a choice of either developing value-added analysis tools or consigning themselves to playing a side game and eventually being acquired. Having just some of the data hasn't proved to be a sustainable strategy.

Rick Wise: You've got to have a relationship with a client – data and information can never do this by themselves. But you've got to back this up with continuously creating higher level analytical techniques that allow you to develop unique insights into a client's circumstances, and that take full advantage of the changing nature of the raw data available. In many instances, we have much more detailed and complete data than we might have expected even a few years ago. At Mercer, for example, we've developed proprietary tools for analysing customer data which are hard for our competitors to replicate. But you've also got to invest continually in the high-level intellectual capital and frameworks which enable you to turn information into practical insights.

Richard Balaban: But starting with the customer is only the first part of the picture. You have to be able to have analytical tools that yield distinctive insights. Back in 1972, Bruce Henderson at The Boston

Consulting Group pretty much invented the idea of strategy consultancy as we now think of it, with techniques such as the growth share matrix and the experience curve. Other consulting firms soon caught on, as did the business schools, which started teaching it to their MBA students. As a result, you've got an entire generation of business graduates going into industry who know how to apply these models: they didn't need consultants to tell them how to use them. But strategy continues to get more complicated, driven by – for example – globalisation, more demanding shareholders, and so on – so you need a new generation of tools to do that. Clients suddenly realise that business planning is something they only do once a year – and even then it's largely viewed as a process to be managed, rather than a potential source of competitive advantage – but that we do all the time, and the whole thing comes full circle.

You also need to be able to apply this knowledge. You can segment the consulting industry along two dimensions – the stakes involved in a particular project (high risk/high return *versus* low risk/low return) and the extent to which the solution is structured or unstructured (Figure 8.3).

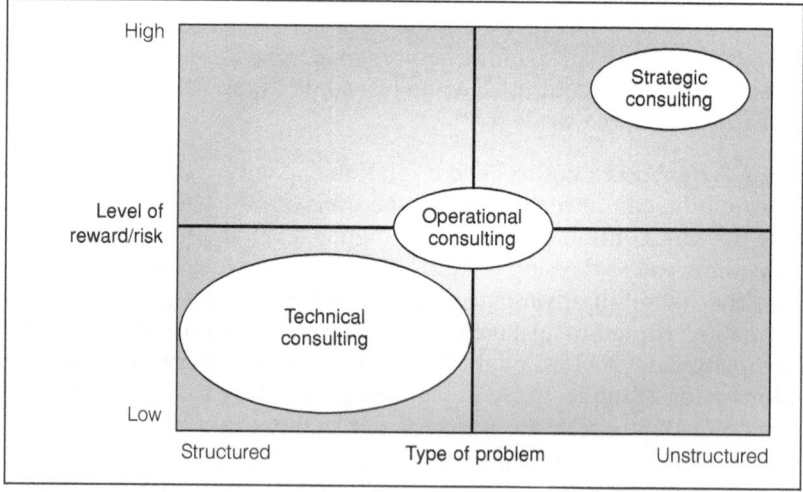

Figure 8.3 *Categorising consulting firms according to the complexity and risk/reward profile of their work*

In the bottom left-hand corner, you've got $100 million systems implementations or projects to optimise your supply chain: these are huge engagements but the skills required are technical. It's about applying an accepted set of rules, and, to be good in this market, you have to be an expert in these rules. In the top right-hand corner, you've got problems which are much more difficult to conceptualise: typically, you're solving things of substantial potential value that no one has solved before. There's no manual to refer to. And, in the middle, you've got operating problems, partly structured, partly unstructured.

Mercer has typically started out in the top right-hand segment, and tended to drive southwest to the limits of our capabilities. Similarly, the operationally-based consultancies, which have typically started in the middle, also drive southwest, towards technical, structured problems. And of course, the originally technical consultants push northeast. The intellectual capital that underpins each of these positions is valuable to a client, in different ways, but what they – clients – really value is the ability to connect these different types of intellectual capital, to act as a bridge, if you will, between strategy and its implementation – and that's something that a consulting firm that's based in the bottom left-hand quadrant finds very difficult to do. It's something that some clients also grapple with: they still subscribe to the notion that strategy is a kind of intellectual dilettanteism, something that's done in a remote part of the organisation with no connection to its day-to-day activities. What matters is the application of business ideas: that's why the client-consultant relationship has changed so much over the last twenty years. ❜

Data, Information and Knowledge: Towards a Value-Based Approach

Two perspectives, then, on the changing intellectual value chain that underpins the consulting industry.

For 'upstream' firms whose focus is primarily on structured, explicit information, the challenge is finding a business model which, contrary to the norms of the consulting industry collates multiple sources of information, not for a single client, but for a group of clients. The current one-to-one approaches are unlikely to satisfy clients in the long-term, given that they can now access a wealth of information which, only a few years ago, was privileged. Yet gathering all the information clients would like is becoming prohibitively expensive if performed on behalf of one client only. The solution has to be for firms in this area to move more towards a structure that combines the scope desired by clients with the cost-effectiveness required by consulting firms, a model not unlike that of a research company, for example. While the last few years have seen research companies try (and largely fail) to enter the consulting market, the next few years may well be marked by consulting firms moving in the opposite direction.

But for high-end strategic consultancies – 'downstream' firms – the challenge lies in distancing themselves from this model – first, by focusing their data gathering activities in certain niche areas, particularly data on their clients' customers; second, by constantly re-inventing their analytical tools in order to produce insights clients could not have perceived for themselves; and, third, perhaps most importantly, doing the one thing that no amount of information aggregation or research could do – bridge the gap between the output of this analysis and the practical steps an organisation needs to take if it's to acquire genuine, sustainable competitive advantage from it.

---

[1] Adrian J Slywotzky, David J Morrison, et al., *Profit Patterns: 30 Ways to Anticipate and Profit from Strategic Forces Re-Shaping Your Business* (New York: Wiley, 1999), pp. 97–124.

[2] John Hagel III and Marc Singer, *Net Worth: Shaping Markets When Customers Make the Rules* (Boston MA: Harvard Business School Press, 1999), pp. 19–20.

9
Structured Methodology:
The Consulting Prisoner's Dilemma

A maturing market means a more sophisticated client base which recognises that generic solutions have little value to add.

Stephen Sprinkle, Global Director of Strategy, Deloitte Consulting

Darrell Rigby, a Vice President at Bain & Company, has studied the use of management tools for the last eight years.

❛ I think the fundamentals of how you succeed in business haven't really changed – companies are trying to create superior results, and you only find superior results at the intersection of a customer opportunity, a competitive vulnerability and a distinctive capability. Seizing any one of those opportunities is hard enough; doing all three together is enormously difficult.

And doing that repeatedly, over years, is an absolutely Herculean task. Executives have trouble keeping their organisations sufficiently motivated on each of those tasks. I also think that there's a tendency in organisations to take a concept and push it a little too far in order to test the boundaries of its effectiveness. When that happens – let me take decentralisation as an example – and you push that too far, organisations start to make mistakes, leaving room for a pundit to come in and argue that decentralisation is killing the organisation and that the solution is centralisation. It's a pendulum: people start touting the weaknesses of decentralisation and

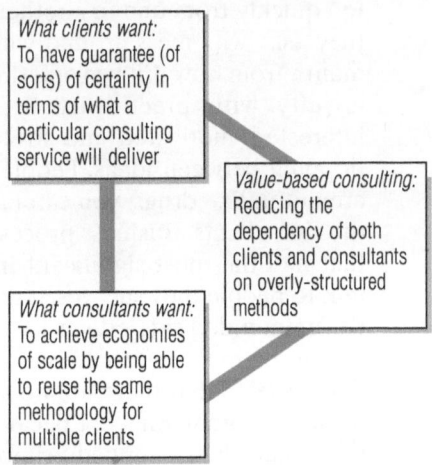

What clients want:
To have guarantee (of sorts) of certainty in terms of what a particular consulting service will deliver

Value-based consulting:
Reducing the dependency of both clients and consultants on overly-structured methods

What consultants want:
To achieve economies of scale by being able to reuse the same methodology for multiple clients

then the momentum shifts in the other direction. These kinds of fads come and go and are actually seen to be a 'quick fix' by many people. Corporate venturing was a popular way for companies to diversify during a time when technology stocks were rising and people were rushing to join in the craze. However, given the current economic climate, the market has come back to a realistic level again. People are recognizing that corporate venturing, no matter how alluring, was not the perfect answer. As a result, they have come back to a single focus on building their core business again. What all that means is that the conditions in a market change, making some tools more or less relevant. Look at scenario planning. It began (in 1994) with usage rates of about 44 per cent worldwide. By the year 2000, this had dropped to 33 per cent even though the satisfaction scores of the people that were using it were quite high. The economic success, the prosperity of the economy, had led people to believe that contingency planning wasn't really necessary, that it was a waste of time. People felt they were doing contingency plans when there would be no contingency. Since the economy turned down in the US and then since the tragedy on 11 September, I've been getting all sorts of calls from reporters as well as companies saying, 'hey we're looking at this thing called "scenario planning"; what can you tell us about scenario planning?' It's not a new tool. Perhaps its benefits are now a little more obvious because scenarios have entered these people's lives that they never imagined were possible.

The other factor that you have to bear in mind when you think about management tools is that organisations can't afford to switch too quickly from one to another, otherwise they reach the point when they ask: 'why mess around with this? We're just wasting our time: a month from now it'll be something else.' So managers have to balance novelty with practicality in order to keep their organisations interested, motivated and stretched – if they're not to end up flip-flopping between ideas. People are learning that management ideas are not unlike drugs: you can take them, but you need to be aware of the side effects. Business process re-engineering was perhaps the tool that had the most significant impact in this respect. Now when you talk to people they say 'yes, it does have benefits but we were blind to the unintended side effects'.

One of my greatest hopes is that the value of our research will demonstrate that there are no silver bullets out there, and that things most often go wrong, not because the tool itself was bad, but because it was naively misapplied. Almost every tool we've ever surveyed has created highly satisfying results for some companies. But, as soon as other companies believe that – because it worked for others, it will also work for them – and don't recognise that management tools never eliminate problems, all they really do is exchange one set of problems for a new set of problems. Unless you go in understanding what those new set of problems are that you're going to be forced to live with, and preparing to deal with those problems from the very beginning, you're likely to be blind-sided and not get nearly as much value out of the tool as you had hoped.

Our research certainly backs up the idea that it's the earlier adopters, who think through how a technique is to be applied to their business, who benefit most. The benefits are less for those who follow, who try and emulate what others have achieved without thinking it through in their own unique context. There's also the factor that anything that can be quickly and easily adopted by one organisation can be just as quickly and easily adopted by everyone, so there's no competitive advantage. The people who have the most success with tools are using the same tools as everybody else, but they're figuring out how to make them work in their organisation, very thoughtfully and very carefully. I don't believe I've ever seen a company lose a competitive position because its competitor adopted a management tool six months or a year earlier than they did. That just doesn't happen.

At the same time, technology can also have a significant impact. There's a classic example of that going on right now in customer relationship management. A lot of firms who began CRM programmes early on tried to set up comprehensive systems with software programs that weren't designed to talk to each other. As a result, they've ended up with a patchwork of very expensive systems that promise extraordinary results but just can't deliver them. Those people are at a disadvantage compared to those who entered the market later when the costs of the technology had declined dramatically. Interoperability has improved tremendously and important lessons have been learned about what you can and cannot do with – for example – data mining. The people who went in with stars in their eyes believing – 'we're going to collect names of all of our customers and then bombard them with messages so they will spend all their available income with us' – are being sorely disappointed.

It's my belief that academics and consultants have a tendency to exaggerate the potential benefits of a management tool and to understate the potential cost to an organisation. Management tools come from two sources. One is a set of tools that is developed in a laboratory or a think-tank. Somebody gets a notion and they become the messianic guru that tries to spread it to the world, irrespective of whether it has or does not have much relevance to the real world. It's just something that the guru believes in strongly.

The second source comes in the form of management tools developed to solve a specific problem at a particular firm. Whoever's involved, whether that's a consulting firm, an academic or a corporation, looks around and says, 'I'll bet there are other companies that have this problem or one very much like it. Let's standardise it, roll it out to our troops, as quickly as possible, and spread it to other companies.' This approach has, I think, the advantage of being standardised but it has the disadvantage of not being customised to the culture and organisational needs of a particular company. I've seen presentations where consultants come in and tell companies that they have to change their management processes to fit in with the consultant's ten step approach. Companies should beware this approach. What they should be saying is: 'you have these management processes, let's see how we should be adapting this tool to fit your management processes.' Any company that tries to change its management processes at the whim of every new tool is going to run itself and its organisation ragged. You just can't do that. You have to pick the direction for your company and then get tools to fit that strategic and cultural direction.

Some of these tools are designed to solve real life problems and they probably can be helpful to other companies. But almost always, the developers of the tool believe that it's an essential tool for far more companies than it truly is. What matters to them is applying the tool ever more quickly, and they tend to overestimate the net benefits to any given organisation. In the long term, this damages the credibility of the management team that foists such tools on their organisation. I don't think it's healthy.

Trying out new techniques is fine, but you have to balance whatever you think the upside is going to be with the potential costs. There's no magic potion or silver bullet. You have to think: 'if it helps, great, then let's go after it'. Executives, in particular, have to champion enduring strategies, not fads. You've got to tie your reputation to something that will last – ambitions like, 'we're going to deliver the best customer service in the industry "or" no-one is going to under-price us'. Then tools can fit into that to help support. Executives should tie their credibility to the strategic concept, not to an individual tool.

Could the whole market for management ideas implode? Certainly.

Absolutely. I think companies get very weary of being bounced from guard-rail to guard-rail. I've talked to organisations that, when the next tool is foisted on them, go through the motions of pretending they're conforming to the principles with the certain knowledge that it'll all go away in six months' time. 'Keep your head down. Don't make too much trouble and it'll pass.'

Familiarity Breeds Contempt

Structured tools methodologies are silver linings that can conceal clouds. It's common sense to say that businesses need disciplined approaches to analysis, change management, systems implementation, and so on: how else would anything get done? Even small children quickly learn that the fastest way to fit the pieces of the jigsaw puzzle together is to find the corners, then all the side pieces, then to start assembling clear, but distinct parts of the picture. They've learnt that they're less likely to finish, more likely to take longer, if they have a completely chaotic approach to assembling the pieces. Entire industries evolve and reach maturation around the same principle. What might start out as an inexperienced small business, producing and selling a new product or service inefficiently, develops into an established corporation with well-oiled manufacture and fulfilment processes.

The same is true for the consulting industry. A new business idea emerges in the market place: consultancies scramble to be the first to bring it to their clients' attention, improvising as they go. After a time, when clients and consultants have become much more familiar with the idea, both will be looking for 'economies of experience' – a faster, more certain way to implement the idea in practice. As *Management Consultancy; What Next?* argued, clients will accept that a consulting firm is 'learning on the job' when a new business idea or technology emerges. Demanding something tried and tested simply isn't feasible and specialist skills are very scarce. But, as the idea or technology becomes more established, clients become less tolerant of guesswork: consulting firms have to train specialist consultants (they have no excuse not to) and develop methodologies that codify their collective experience in order to reduce the risk of failure. Thus, during the life of a business idea, the value perceived by clients switches, from fresh thinking and having bright consultants willing and able to work out how a new idea can be applied in a given context, to wanting trained resources and an established way of doing something (Figure 9.1).

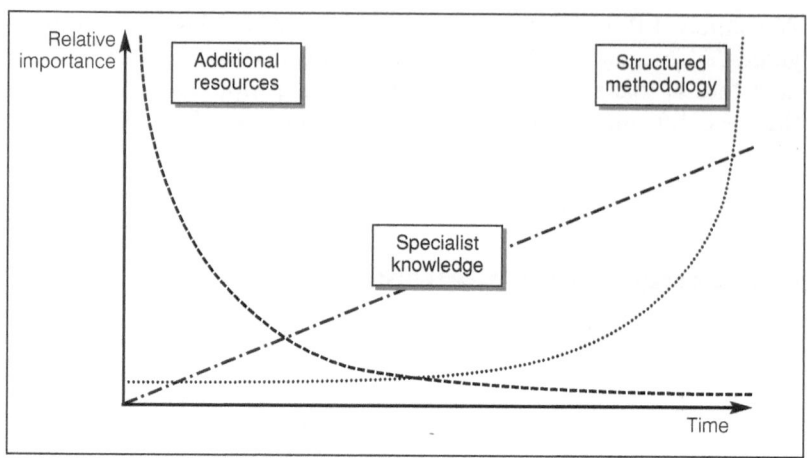

Figure 9.1 *Changing perceptions of value as a business idea matures*

The 'cloud' lies in the fact that this evolution is accompanied by another more malign one. As a set way of doing something emerges, people's attention shifts, from the end to the means: they become less concerned with what they are trying to achieve, and spend more time on how they're going to achieve it. We've all seen it happen. The managers who focus on filling in all the appraisal forms they're supposed to, rather than seeing whether the performance of their staff has improved. The PR company that concentrates on getting a specified number of mentions for their client, and doesn't analyse whether these have actually had an impact on the latter's name recognition.

Of course, these attitudes are largely driven by performance measurements. It's much easier to measure a process than its outcome: processes can be decomposed into discrete elements and people can be made accountable for them. By contrast, outcomes are affected by many different variables. The behaviour of people is determined by their values, their home life, their customers, probably far more than their managers. Name recognition is as much determined by the sector your business is in, your market share, the length of time you've been established, and so on: it's very difficult to design, let alone calibrate, a model that could measure all these different factors. Such behaviour is driven bottom-up within organisations, as well as top-down: people want to be measured in process terms because this is what they can control. What else is a process but a definition of the breadth of control an individual (or group of individuals) has? People can perform processes much more easily than they can deliver results. Even senior executives, more likely to be measured by their outputs than the people who work for them, can be tempted to rest their performance on the process – 'doing the job well' – especially when market conditions and other factors are difficult.

Applied to the consulting industry, there's academic evidence to suggest that you can become too 'structured' in your approach. Management theory work best when it's bones without much flesh – good ideas that need to be customised to fit the unique needs of a particular organisation. Once an idea has matured into a bandwagon, people tend to think there is only one way to apply it, that there are a set of rules they have to follow, that their organisation has to change to fit the idea, not *vice versa*. Moreover, the companies that 'buy' the more mature version of the idea are less likely to achieve substantial benefits than their earlier adopting rivals, partly because they're buying for the wrong reasons (because everyone else has bought it) and partly because they're working on the basis that they have to apply a prescribed set of rules rather than think through how the idea could be made to work. A vicious circle emerges: the early adopters – who have to customise the theory to their individual practice – obtain substantial benefits; these, in turn, encourage companies who had not initially believed that they would obtain a significant advantage from the idea, to adopt it, effectively against their better judgement. These companies look for consultants with a track-record: next to client reference sites, the next most effective way of doing this is to have a formal process. One irony here is that the consultancies that don't have too codified an approach may well be those able to add the greatest value, but they are also the firms least likely to find themselves on a short-list for work.

Those early adopter benchmarks also create a standard which drives the consulting industry: consultants, too, will be looking to replicate what their first clients achieved, rather than reinvent the idea on behalf of each new client. As the approach becomes more rigid, the level of customisation goes down, as do the benefits achieved – but people's reaction is to demand greater structure because they see it as a means of increasing the likelihood that the idea will be made to deliver the anticipated benefits. Another irony of this situation is that it's quite the reverse: it's lack of structure that produces the most benefits.

But costs also play a part here. Inventing how an idea is applied to a particular organisation is expensive, both for the client (in terms of fees) and the consulting company (it's difficult to utilise resources effectively when you're not sure what you need when). It therefore makes sense on both sides for a structured methodology to play some role: it may limit the ultimate benefits in the long-term, but – crucially – it also contains the visible outlays in the short-term (Figure 9.2).

When a new management idea first appears, both the benefits and costs are high (1 in Figure 9.2). As a way of applying the idea starts to be established, the costs fall dramatically, and the difference between the benefits and costs is at its highest (2). Finally, as the application methodology becomes more rigid, there comes a point (3) when the costs actually exceed the benefits. It's a pattern we've seen over and over again: business process re-engineering, enterprise resource planning, and so on. Current bandwagons – customer relationship management, for example – may well go the same way.

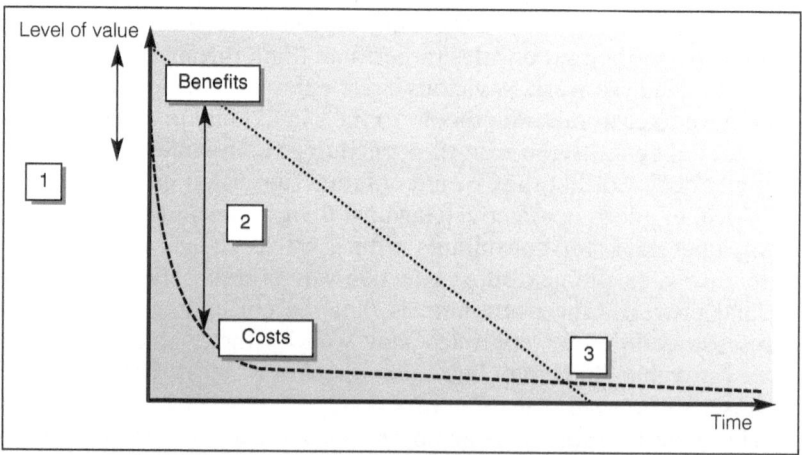

Figure 9.2 *The relative benefits and costs of a management idea*

A further factor is that measuring the output of consulting assignments is also notoriously hard: not only must you take into account the range of variables which would affect an internal employee, but you have to add to them all variables of the client-consultant relationship. Clients may choose not to implement what consultants suggest; or, if they do, they may not do so effectively.

Consulting firms have largely resisted the pressure from clients, to be measured and paid in relation to ultimate output (increases in sales or decreases in profits), and have preferred to be evaluated against process benchmarks – timely delivery, the completion of a roll-out programme.

Structured methodologies are therefore something that clients want, especially as an idea becomes more familiar; they're also something that consulting firms are happy to develop, as they make their costs more controllable and their role more measurable. The trouble is that they can become too 'successful' for their own good, actually destroying the value they're intended to create. So it's a balance: you need some methodology, but not too much. Saying it is easy – what's difficult in practice is that the internal and external pressure to 'methodologise' is almost irresistible. 'Clients often don't believe you can do something unless you've got a manual the size of a telephone directory', said one partner I talked to. 'However much you want to be able to say "look, I'm not sure how this will work in your case, but we can work it out" you know that's not what they want to hear. Clients buy certainty.' Strategy firms have more room to manoeuvre, but even they're not exempt from such pressures. As one strategy consultant put it: 'The first year of the e-business boom [1999] was great. For the first time, you really felt as though clients were hiring you for the right reason – to think the unthinkable. But as the e-business bandwagon gathered momentum, you found more clients saying "I want one of those". They wanted to be sure they could have something up and running in next to no time because they thought they'd otherwise be missing the boat. It was the equivalent of going from having a house designed for you to buying one already built: the only changes you could make were cosmetic; you weren't allowed to challenge anything.'

Consultants are also aware of the internal pressures: 'In theory', said one, 'if you're measured on fees earned, then working inefficiently shouldn't be a problem – as long as the client's willing to fund it. And it isn't a problem if you think of each project as a separate activity. In reality, however, a consulting firm is the sum of all of these projects, and inefficiencies in one area cause problems in others. If I'm running one of these amoeba-like engagements, and I want a specialist in a particular area for a couple of days, but I'm not sure when, what do I do? Do I hire – at extra expense – an outside contractor? Do I find someone internally and prevent them from being 100 per cent occupied on another project? You wouldn't try and run a machine in which the different cogs couldn't work together.'

Structured Methodology: Towards a Value-Based Approach

Is there a way of meeting the demand for efficiency and certainty without giving in to standardisation and rigidity? The key has to lie in breaking the link between certainty and efficiency.

Clients need to understand that efficiency doesn't necessarily guarantee certainty. Just because there is a process for doing something, doesn't mean that the outcome of that process can be assured. In fact, as we're already seen, efficiency can actually be a barrier to delivering value. But it's tempting for a client, looking at an efficiently run process, to believe that it is more likely to yield benefits than an inefficiently run one. This is partly what we want to believe – 'if I hire these consultants who have a ten-step plan for doing what I want, I'm more likely to get what I want'. It's partly what we're trained to believe – 'if I run my business more efficiently, it will be more successful'. And consultants pander to this assumption, by packing their presentations and proposals with structured methods. It's also an example of the prisoner's dilemma (Figure 9.3). If two consulting firms (A and B) are pitching for the same business, and neither offers a standardised methodology, then the client ultimately benefits, because a more customised approach is more likely to yield the required results. If Consulting Firm A unveils a standardised approach

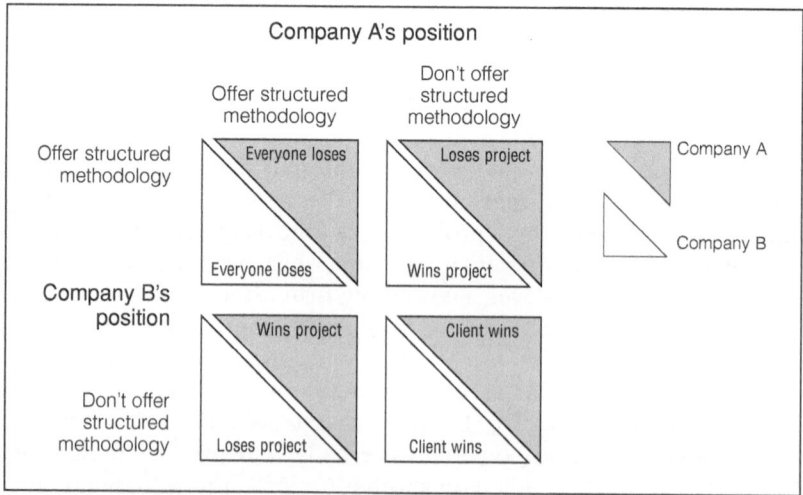

Figure 9.3 *The consulting 'prisoner's dilemma'*

that appears to offer the client a greater chance of success, then it is more likely to win the work, although the benefits the client realises will probably be reduced. If both Firms A and B present a standardised process, then everyone loses: the client, because such a process is likely to cut into the anticipated benefits of the work; and the consulting firms, because they end up competing on the efficiency with which they can produce these benefits (that is, price).

For both Firms A and B, it's tempting to play the structured methodology card, because it's likely to win them the client's business. Only by acting together, can they create a situation in which the client wins, where the firms compete on the quality of their solution, not on its price.

10

Networking: Adding Value in a Joined Up World

Individual dot.coms may have come and gone, but their legacy of crumbling boundaries between industries and organisations is still with us. One client I spoke to summed it up: 'ten years ago, if we'd spotted a new business opportunity – launching a new product, entering a new market – the chances are that it would be something we'd pursue independently. Today, we'd almost immediately be looking for partners. Why go through the effort (and time) or establishing your brand in a new market, if you can piggy-back off someone else's? Why spend money (and time) developing a new proprietary product, when someone else may have the technology you need albeit applied in a different context?'

> **What clients want:** Access to the 'reach' of a consultant's network, without ceding power

> **Value-based consulting:** Clients and consultancies working together to create new business ventures

> **What consultants want:** To play a leading role in creating new businesses and markets. To be seen as movers and shakers, not just observers

But, when it comes to tracking down new partners, especially those in different industries or geographies, even large corporations can find themselves at a loss. 'The deals we're now putting together might involve a telecommunications company and a travel agency,' said the finance director of one financial institution, 'but the only people I tend to talk to are finance directors in other institutions.' The economy might be becoming networked, but individuals aren't necessarily.

It follows that one of the most recent ways in which consultants have been able to add value to clients is to extend the latter's 'reach' – to open doors for them in other companies, in other sectors which they could not open for themselves, or only with difficulty. 'If we're helping an organisation launch a new business', said one consultant, 'they don't look to us any more just to design the processes or implement the systems. They want our input into identifying suitable partners and facilitating the relationship.' 'It was great', recalls a client, 'when we needed to negotiate deals with other companies, the consultants we were working with simply got on the phone to their friends.'

It's a role that puts consultants more on a par with investment bankers, but there's a significant difference. Clients are accustomed to investment bankers charging by transaction – bringing potentially interested parties together, brokering the deal, advising on the ownership structure of the new company, and so on. For consultants, it's quite different: clients are used to paying for the time expended, not for the value of the deal or, indeed, the network of contacts the consulting firm has built up. From the client's point of view, asking consultants to 'make a few phone calls' on their behalf sounds exactly like the kind of value-added service that consultants should be doing to justify their exorbitant rates. Moreover, that's exactly how consulting firms viewed this role until they began to appreciate its potential value. Switching now – to something more like an investment bank's approach to charging – will therefore be a particularly noticeable change.

But that's what consultants want – and need – to do. One consultant I talked to recalled working for a manufacturing company that was looking to improve the data it had on its customers. 'Like other manufacturers, this client found itself in a position where its distributors knew far more about the final consumers of its products than it did itself. If it was going to respond more effectively to changing needs, then it was going to have to develop its own sources of market information and, rather than go down the conventional route of commissioning focus groups, it wanted to launch a direct survey, via the relevant press. It was an expensive route, and, to spread the cost, the company wanted to find others willing to fund questions of their own. As a consulting firm, we were hired to help with the mechanics of the survey – how to process the data, what kind of analysis should be carried out, how the results could be assimilated back into the client's product development process. But what the client really needed was help in identifying partners to spread the costs of the survey and subsequent data processing. We were able to put them in touch with half a dozen other companies, in different sectors, all of whom bought into the survey. It was a great success from the client's viewpoint, but rather less from our own. Although we were delighted to have made a significant contribution to the success of the project – indeed, I honestly don't think it could have happened without us – we only got paid for the hours we put in, not for the value we created. There was no recognition that those six contacts represented years of accumulated intellectual capital.'

But concerns over pricing point to a much more fundamental conundrum, and one that challenges the balance of power in the client-consulting relationship. If, as the example just quoted suggests, a consulting firm's ability to field a variety of potential partners does play a fundamental role in developing new business opportunities, then clients will be more dependent, not so much on 'generic' consultancy, but on a particular consulting firm. Of course, a client hiring a consulting firm to implement a system to support a new business is dependent on that consultancy doing its job properly, but if it doesn't, then the client has the option of being able to switch to alternative consultancy, capable of supplying a comparable service. The difference between this, and the role that a consulting company plays in finding and bringing together potential partners, is that every consulting firms' network is unique. If something goes wrong, it's much less easy to bring in another firm to pick up the pieces – you have to start again. When you – the client – hire a consulting firm to implement a system, what you're doing is bringing that firm into your network of suppliers and partners. But when you ask a consulting firm to tap into its network on your behalf, you become part of the consulting firm's network. The hub of the network – power – has shifted from you to the consulting firm.

It seems to me that the concerns over charging fees for networking are much more to do with this issue, than simple economics. Consulting companies want the importance of their role acknowledged – and fees are one way of doing this; clients are reluctant to elevate consultants beyond the current position. The irony here is that being able to facilitate networks and partnering arrangements is something that consulting firms are uniquely well-positioned to do. The industry/geography range of their networks is far wider than that to which most clients have access. Moreover, their diversity of contacts in different business functions (operational as well as strategic) is far greater than that to which even most investment banks can claim. But unlocking the considerable value that consulting firms should be able to add depends on finding an equilibrium in the balance of power between clients and consultants.

And, for all the considerable lip service paid to the idea of clients and consultants working 'in partnership' together, it's rare that this genuinely happens in practice.

In the conventional way of thinking about consultancy, partnership working, like power, is a function of knowledge. When a 'new' consulting service hits the market, consultants have the upper hand: they'll typically have invested in researching it, building up their track record, developing a structured approach to deliver; clients, by contrast, may have heard very little about it, except perhaps brief highlights culled from the business press. During the course of the project, as the client learns more, the balance of power shifts away from the consultant to the client (Figure 10.1a).

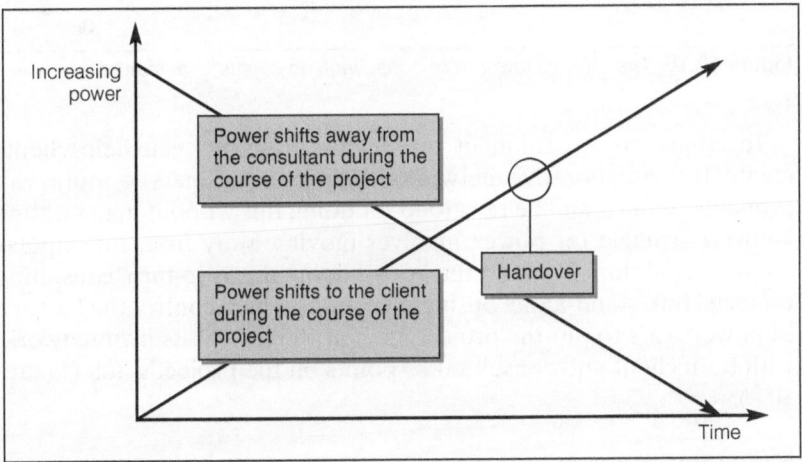

Figure 10.1a *The shift in power from consultants to clients in a 'new' service*

By the time that a formal handover occurs, the client (ideally) knows as much as the consulting firm on the subject in question and is capable of taking complete ownership of the project: the transfer of power is complete. As a consulting service 'matures' – as clients have a greater understanding about it prior to hiring a consulting firm – the knowledge gap between consultant and client clearly shrinks (Figure 10.1b). Handover may still occur at the same point, but the client will have been 'in control' for much longer. Ultimately, the initial 'power gap' between clients and consultants disappears – along with the market for that particular service.

Value-Based Consulting

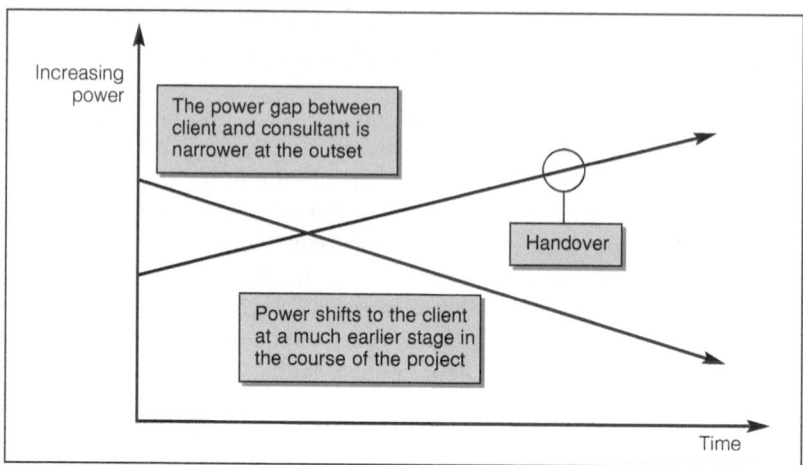

Figure 10.1b *The shift in power from consultants to clients in a 'mature' service*

To create an environment in which consultants can help clients create the collaborative networks their new business venture will probably require, and be rewarded for doing this without inaugurating a covert struggle for power, involves moving away from this bipolar conventional thinking. And the key to doing this is to turn 'consulting projects' into stand-alone business ventures which control the balance of power – it's to put the project itself at the hub of its own network, with both client and consultant as points on the project's hub (Figure 10.2).

Proponix: Banking on Collaboration

There aren't so many examples of this in practice, but one such is Proponix, a joint venture between the Australia and New Zealand Banking Group, Barclays and Bank of Montreal, together with American Management Systems.

Trade finance is a profitable business that often plays a crucial role in building long-term client relationships. But the back-office processing of trade transactions – letters of credit and international collections – is a high-cost, low return business. For most financial institutions it's Catch-22: they depend on the processing systems, but can't justify investing in them because the actual value they generate is negligible.

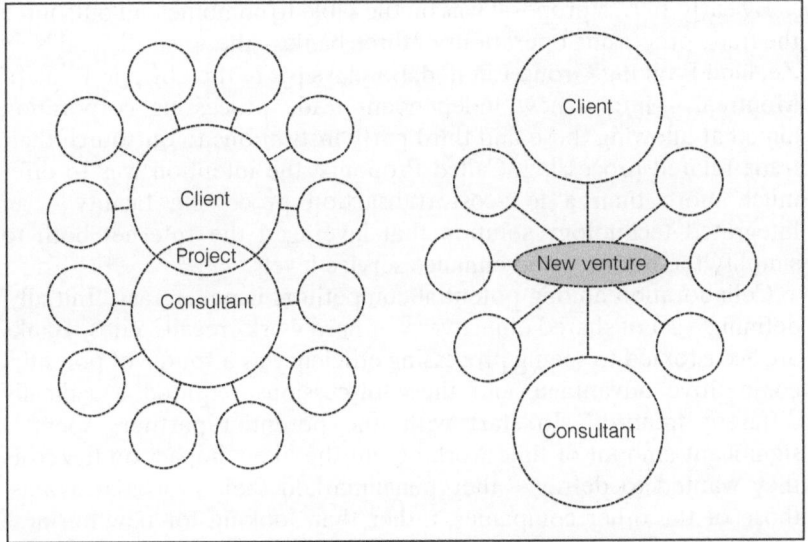

Figure 10.2 *Putting the consulting project at the centre of the network hub*

‘ We did a study for one bank in Australia’, says David Yates, the Chief Operating Officer, Europe, ‘which concluded that there was no way a rational businessman would spend more money on these systems. But, at the same time, a similar study for another bank showed that chronic lack of investment was driving the bank into a corner. Either they’d have to invest eventually (otherwise the system would collapse), even though to do so went against almost every criterion they had for IT investment, or they would have to exit the market – an unacceptable strategy in most cases because trade finance tends to be so heavily interlinked with other lines of business that a bank would risk losing up to 40 per cent of its corporate business by not offering trade finance. Or they could outsource the work. In fact, outsourcing was the only viable option, but the only companies offering this kind of service were other banks focused on reducing their costs, rather than offering a service. But there was an alternative – at least in theory. If you could build up sufficient scale in trade processing, if you could industrialise it, in effect – and that was something no bank could do independently so it would necessarily involve bringing competing banks together – then you could have a sustainable business in its own right, and one that might be worth investing in.’

By early 1999, a proposal was on the table to combine and outsource the trade processing operations of three banks – the Australia and New Zealand Banking Group Limited, Barclays plc in the UK and Bank of Montreal – into a new, independent trade processing corporation aimed at allowing these and third party institutions to outsource their trade-related processing. Called Proponix, the intention was to offer much more than a low-cost transaction processing facility – an integrated technology solution that leveraged the Internet both to simplify the business and enhance service levels.

Collaboration among potential competitors is never easy. 'Initially, defining a set of shared objectives was hard work', recalls Yates. 'Banks are accustomed to seeing processing efficiency as a source of potential competitive advantage, and these discussions required a radically different mindset'. To start with, the potential partners spent a significant amount of time working out the likely impact on the costs they wanted to defray – they benchmarked their processes against those of the other companies, rather than looking for new business opportunities. In the end, it was the financial extrapolations – the idea of creating a new business focused on generating revenue for all those involved, rather than just cutting the costs of the individual partners – which brought everyone together. But the negotiations didn't stop there: even once the financial arrangement for the initial participants had been agreed, there was a complex pricing structure which had to be defined for future customers. According to Yates, 'that was something which meant that – really for the first time – we had to step outside our roles as representatives of our respective organisations and think about things from the perspective of the new business. It made us appreciate that we had to be absolutely clear, when we said something, when we did something, the role we were playing. Were we representing the interests of our own companies? Were we representing the interests of the new one? Confusing them would be fatal: we'd have completely lost the plot.'

Another important factor was the decision to set up an interim management team even before the small print on the contracts had been agreed. 'If we'd sat around waiting for the lawyers to finalise everything, it's quite likely that someone else would have entered the market'. By early summer 2001, Proponix's Board, its Chief Executive, Bill Graham, and its executive management team were in place. The focus now shifted from setting up basic business functions and appointing personnel to completing the final stages of transition, from a project run by the three banks and AMS, to a stand-alone business.

This took two forms. Internally, a schedule for the transition of each component had to be drawn up. Externally, a marketing process had to be started to provide the venture with a public profile. 'Keeping a sense of continuity has been really important,' says Graham, 'although we were creating a new entity, it was important to preserve the underlying thought processes that had gone on. Everything depends on our being able to execute the plans that had been drawn up, and we have to ensure that the people involved retained a strong sense of ownership. One of the advantages of bringing three institutions like these together, and going through a very detailed series of negotiations with them, is that the accumulated body of knowledge we have about this business will be a real differentiator in the marketplace. A priority now is to ensure that none of that knowledge is lost, and that, as we acquire customers, we can add their knowledge to our own.'

Graham, whose background is in large-scale mergers and acquisitions in the financial services sector, sees deals like Proponix revolving around three issues: capital, technology and people. 'By the time I came on board, the first two of these were pretty much sorted. The key variable, which will determine the success or failure of the venture, is people.' What kind of characteristics matter? 'You've got to have good teamwork. The complexity of deals like this is such that you need people with very specific skill sets – general management acumen just isn't enough – and that means that you've got to be sure that they can all work together. No one person can do everything, so it has to be an integrated effort.' This also means that the organisation has to be structured around collaborative decision-making rather than hierarchical control. And the global nature of the business has been an additional challenge, Graham points out. 'We have to be clear about how an inherently complex business model will work in different continents, and we've got to design an organisation capable of managing it at all levels. We therefore made a conscious decision to build an international executive.'

Getting people to join from the participating banks hasn't been a problem. According to Graham, 90 per cent of the people involved in the in-house trade processing teams want to move to the new company. 'It makes complete sense', he says. 'Why stay in an organisation where you're only one small part of what it does, when you can go somewhere where it's the core business, where your knowledge and skills will be valued and developed.' But Proponix has to balance the continuity this gives it, with the need to create a new, more service-orientated organisation. People coming from the banks are not transferred from the old company to the new (as they typically are in outsourcing deals). Everyone has to apply for a new job and resign from the old.

Another important factor, from Graham's perspective, is managing expectations. 'We're in the middle of a transition from the partners making decisions jointly to an environment in which Proponix – a third-party, in effect – will be making these decisions on their behalf. Our shareholders are also our clients, and we have to distinguish between these roles very clearly, otherwise it's a recipe for confusion about what we're trying to do.' External expectations also have to be managed. Even from the outset, Proponix's marketing generated considerable interest in the financial services community. 'With the combined business of the banks, we'll instantly be one of the largest trade finance processing organisations in the world', says Graham. 'We're also addressing a problem common to almost every financial institution with what should be a win-win solution. The high volume of transactions we will be processing justifies a level of investment and service which few institutions could afford independently.' He believes the fact that Proponix was created, not by a group of people with an idea, but in response to a very specific problem shared by many organisations, produces a high level of empathy between Proponix itself and its potential customers.

The attitude of the three banks has been fundamental in getting Proponix off the ground. According to Graham: 'We know of at least one other bank that had this idea, but when it got to actually doing something, came up with a list of 15 reasons why they shouldn't do anything. By contrast, both Barclays and the Australia and New Zealand Banking Group Limited already had substantial trade processing facilities, the success of which meant they had considerable internal credibility. Barclays and Bank of Montreal had experience in outsourcing operations and in creating spin-off companies. And all three institutions had a strong management style, willing to make a commitment and stick with it.' Other partners had been considered but dropped out, either because they lacked the skills required, or because there was insufficient sponsorship from senior management.

But perhaps the most significant factor in determining whether an institution would make an effective partner was the extent to which it had gone through what Graham terms 'the disaggregation process.' 'Financial institutions, he says, 'have been accustomed to a level playing field. They knew who the competition was and they could take steps internally to meet whatever competitive threats emerged. But that's changing. Banks have to start asking very fundamental questions. Who are we? What are our competitive strengths? What do we need? What do we not need? Rather than look in-house for the answer, they're turning to other companies to create a joint solution, so we're seeing more spin-off companies, joint ventures, strategic alliances. Organisations are disaggregating themselves, and reforming around providing solutions. In this kind of environment, it becomes very difficult to provide a clear career path to those involved: why should anyone invest in training themselves or their staff if they can't be confident that the skills acquired will be made use of? It's much better to take functions which require a high degree of technical skill and establish them as stand-alone businesses, able to focus on this type of work. Essentially, Proponix has come into being because the three partner banks all thought along these lines and, to some extent, were going through the process of disaggregating themselves. If they'd still been clinging to the traditional, monolithic approach to financial services, it couldn't have worked.' At AMS, David Yates agrees. 'We'd had discussions with several other banks in a similar vein, but they'd all been put off by the complexity of what was involved, and the fact that no one had done anything quite like this before. What struck us about the current partners was the commitment they had at the highest levels to take this plunge.'

What was different because AMS was in the mix? Like the banks themselves, Graham sees the firm as a 'solutions provider', willing to pluck components from different organisations and bring them together for a common person, rather than mired in the conventional approach to consulting projects. But, in his view, it's not just that the consulting company has been comfortable working in a disaggregated environment that's important. 'It's the skills it breeds. Planning, for example, is hugely important because this isn't so much one project as a multitude of small, but highly-interdependent projects. Some of the other consultancies I've dealt with have become bogged down in the sheer complexity of what's involved.'

It was also important to have a non-banking partner in the mix because they brought different skills, in particular the ability to facilitate a decision-making process. 'AMS's role', says Graham, 'went well beyond what we typically think of as a consultancy: it played a critical role resolving potential conflict between the other partners'. 'Any kind of arrangement like this needs an outsider', says Yates, 'if you all work in the same sector there's, frankly, a danger that you might end up with unrealistic expectations and misunderstood objectives'.

Trust, Yates believes, has been one of the most significant critical success factors. 'The risk/reward profile was different for us', says Yates. 'We were the prime beneficiary of the contracts Proponix was awarding. Integrating the three banks' systems was complicated stuff: obviously, a lot depended on our getting it right, and so we took on a greater proportion of the risk than we would normally accept in a standard commercial contract. But we were also the first to see a reward, in that we were being paid for this work, whereas the other partners were waiting for the venture to start reducing their costs by outsourcing and generating an income through third-party business. It helped that we knew each other as people and organisations before this started. AMS has also played an important facilitative role. 'Historically', says Yates, 'banks have tended to make bilateral alliances, but this kind of venture requires, not only more than two partners in order to acquire the critical mass to convince the markets, but a plethora of additional suppliers. They aren't accustomed to thinking about inter-company communication – their focus has always been on their own business. By contrast, taking ideas to different companies and different industries is one of the core capabilities of a consulting firm. The consultant can therefore play an important role in persuading traditional competitors that they may, in fact, create more value through collaboration.

'We aren't an equal partner in the business', adds Yates, 'but it's very different to a conventional supplier/customer relationship. For one thing, we had to act in the spirit of the agreement, rather than according to what it might or might not have said in the detail of the contract. Because we'd been involved in the earliest discussions, we shared a common set of fundamental principles with the other partners – a code of conduct, if you like, that we all had to stick to. If we'd allowed anyone to break the rules in the early days, then what was still a comparatively fragile edifice could have been destroyed. But we've also had to overcome the traditional "them and us" tension that bedevils so many consulting relationships. We've really been able to see the consulting industry from the client's perspective, and that's made us much more conscious of some issues internal to the consulting market. Because we're now the budget holders, we find ourselves being extremely clear about precisely what we looking for from a consulting firm – and paying for that and nothing more. We don't want to buy generic consultancy: we want the "best of breed", and that's something that no one firm can guarantee in all areas.'

Networking: Towards a Value-Based Approach

What can ventures like Proponix teach us about the emerging role of consulting firms as network facilitators, perhaps even network creators?

In the introduction to this book, I talked about the richness, reach and affiliation of consulting firms (all characteristics which Philip Evans and Thomas Wurster, in *Blown to Bits*[1], identified as being key to the successful businesses in the 'networked' economy). Of these three, I argued that richness (specialist knowledge) is an area where clients and (good) consultancies are roughly equivalent, but that consultants had an enormous advantage when it came to reach (good consultants know more people in more industries), and this placed consulting firms in pole position to become the market-makers of the future. What was lacking, I suggested, was 'affiliation' – the ability and willingness of consulting firms to ally themselves with certain industries or client groupings, to overcome the traditional supplier-customer relationship – and, indeed, all the lip-service paid to the idea of partnership working – and put themselves in their clients' shoes. This wasn't a cultural reluctance, so much as commercial sense: consulting firms have conventionally steered clear of putting their eggs in a small number of baskets, preferring to spread them, portfolio-like, between emerging and mature markets.

With Proponix, American Management Systems has gone some way towards overcoming this inhibition – and it's clearly yielding dividends. It's exploited its richness of knowledge in the financial services technology arena; it's used its reach both to facilitate the relationship between potentially competing institutions as well as approach potential customers; finally, it's affiliated itself with the banking partners – even to the point of buying consultancy on their behalf. If the firm had exhibited only one of these characteristics, it's unlikely that the venture would have progressed so well. You have to do all three.

---

[1] Philip Evans and Thomas S Wurster, *Blown to Bits: How the New Economics of Information Transforms Strategy* (Harvard Business School Press, 2000).

11

Change Management: Do Consultants Have a Legacy?

The job of the chief executive is getting lonelier and the pressure to produce results is getting greater.

John Donahoe, Worldwide Managing Director, Bain & Co

What difference do consultants actually make? Would business today look very different if consulting hadn't been invented? Essentially, all clients buy consultants because they want to change something – it may be their strategy, the way they're planning to market a new product, their organisation or their people. As we've noted in the last few chapters, an organisation may lack the resources and/or the specialist skills to deal with a problem. It may not know what to do, or have no clear idea how to do what it wants to do. It may require more information before it can make a decision, or it may need help managing potential partners.

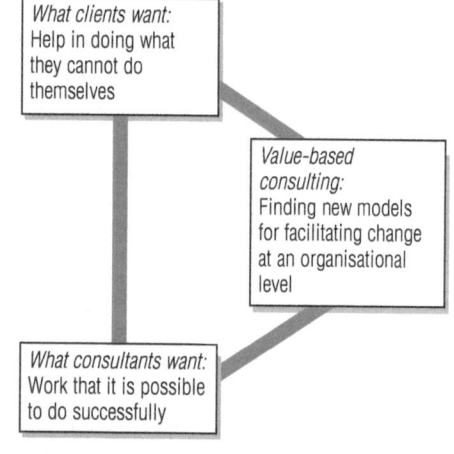

What clients want:
Help in doing what they cannot do themselves

Value-based consulting:
Finding new models for facilitating change at an organisational level

What consultants want:
Work that it is possible to do successfully

This is an issue that has come into clearer focus in recent years, as a result of the 1998–2000 e-business boom. The short length of time it took for the enthusiasm with e-business to peak – and then trough – is a testimony both to the speed with which management ideas – like other sources of information – now flow around the economy and also business people's apparently insatiable desire for a universal panacea. Under pressure from so many sources and despite so many past disappointments, there are few people – whether they are chief executives or middle managers – who can resist grasping at every new idea that emerges, in the hope that it provides the longed-for solution that has, somehow, evaded all of their competitors. As the eighteenth century poet Alexander Pope succinctly put it: 'Hope springs eternal in the human breast, / Man never is, but to be, blest.' The 'bigger' the management idea – the number of its apologists, the length of newspaper columns describing it, the value of benefits associated with it – the greater the level of hope that accompanies it. The evangelists for e-business talk of it changing the very economic rules on which businesses are grounded: it's hardly surprising then, that the levels of hope associated with it were astronomically, and unprecedentedly, high. According to one executive of a large ('old' economy) corporation: 'the key effect e-business had on us, was to change our perception of ourselves. Suddenly, we could do things that we hadn't done before – whether that was entering a new market in record time, or slimming down our processes to a point that we couldn't have envisaged even five years ago. E-business seemed to make the impossible, possible.' And consultants played a role in this – providing resources in the early days when it was proving difficult to free up the client's own staff from their day jobs; injecting momentum into complex projects; providing a road-map during periods when the competitive landscape seemed to be shifting almost daily. 'These projects were much more complex than you tended to imagine at first', commented another client, 'and significantly more so than most "conventional" business ventures. You were faced with many important decisions, almost all of which have to be made within a very short space of time – What kind of hardware should you use? Will this software vendor still be around in a year or so's time? How do I manage security? You were juggling many more – and more interrelated – balls in the air at the same time. You can't have a high-level plan and assume that the "usual" tasks will need to take place: you have to work at a much lower level. And because there are so many tasks that need to be completed, the only possible way you can map the project as a whole is in relation to outcomes, not processes – and this is something that's quite alien to a lot of managers.' The ability of consultants to manage such complexity earned them a reputation – during the e-business boom – of being integral to that process of making visions a reality, but it's also earned them – in the aftermath of that boom – the reputation of not making good on their promises. Indeed, if you look around the consulting market today, you find plenty of clients who say they are less willing to work with (what they perceive to be) the largely US-based consulting firms (new and old) which, they believe, were instrumental in peddling unviable e-business ideas. Looking for shorter projects, with more definite paybacks and which can be implemented, many of them talk about working with smaller, lower-cost, national firms which were not caught up in the e-business hype.

Such adjustments in client buying behaviour may be temporary: a more likely long-term effect is that the bar of expectations from consultants has been significantly raised. It may be that consultants were promising what they could not deliver, but clients continue to look back on this period with nostalgia. 'Consultancy', said one executive I talked to, 'should be about taking your business forward – often in small ways, sometimes in big ones. In the late 1990s, there was a brief moment when that seemed to be the case. We, as clients, were undoubtedly being naïve, but because we wanted to be naïve, just as we wanted it to be true. And we don't want, now, to believe that it's not.' 'We're hiring consultants to cut costs', said another. 'We know that's what we have to do, and we're getting on and doing it. But it's so depressing. None of us dares admit it, but I think we all hanker after the days in which new ideas and possibilities seemed to emerge at every meeting.'

The consultants, of course, are caught between a rock and a hard place. They want to be committed only to those projects which they can deliver (after all, they have, in many cases, seriously damaged reputations to salvage), but they're being judged – subliminally at least – by very different standards – the standards, indeed, of a different age.

The False Promise of Change Management

Consulting firms change things in two ways, either by helping other people to do it, or by doing it themselves.

Historically, it's the first of these approaches that has been predominant: the role of the consultant has been to support, advise, and facilitate. But, as clients have demanded more certain and speedier results, an increasing number of consulting firms have switched to the second. Consultants have become managers, doers, outsourcers. The underlying difference between the two approaches comes down to the 'thing' that the consultant is trying to change.

Where the consultant is playing an advisory role, it's the individuals with whom the consultants work who are supposed to change. A consultant supporting an executive in a new role is supposed to coach the executive in the skills he or she will need. A consultant who facilitates a strategic review should be showing those involved from the client side how to do it for themselves in the future. By contrast, the aim of bringing in a consultant to do something is to change the organisation as a whole – to put in a new system, enter a new market, cut its costs. Whereas the impact of the consultant-as-facilitator at the organisational level is primarily indirect – mediated through the people with whom the consultant works – the impact of the consultant-as-manager is direct. Furthermore, the consultant-as-manager has only an indirect effect on the people in the client's organisation (Figure 11.1).

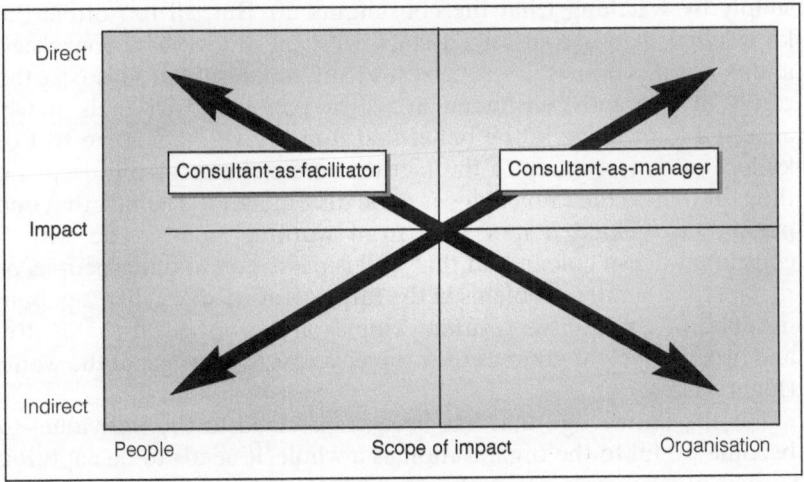

Figure 11.1 *The two change models of consulting*

The nature of the impact is very different in each case. Crucially, the consultant-as-facilitator would expect to have a long-term impact on the intellectual capital of an organisation: he or she will have taught the client's staff a new skill, theoretically enabling them to resolve similar issues which arise in the future. The consultant-as-manager's impact is more immediate – a call centre may be opened or a system implemented. The skills required to do this may have only short-term value to the client's organisation – in operations management terms, they are the skills required to set up a new process, not to run it. There may be little point in an organisation acquiring such skills. Typically, the impact of a consultant-as-facilitator is incremental: that of a consultant-as-manager involves a step change. The level of knowledge transfer from consultant to client should therefore be much higher where the consultant is a facilitator.

If you ask clients about what they really want and the extent to which consultants meet their expectations, the response is usually along the lines of (as one client put it): 'we don't want to become dependent on the consultants involved. They're an expensive resource and we want to get the maximum value from it. The only way of doing that is to ensure that there is some level of skills transfer, but that's something which is difficult to achieve, and even harder to measure.' Typically, clients think they do learn from working with consultants – whether this is via a formal process of transferring knowledge or simply by watching what the consultants do. But, all too often, the knowledge that is transferred stays with the individuals concerned and is not disseminated to the rest of the organisation. This isn't the result of a grand conspiracy in which people hoard their newly acquired knowledge in the belief that it may give them some sort of kudos, but is testimony to the fact that most clients give little or no thought to how the knowledge is to be disseminated. The fact that one person has gained a new skill from working side-by-side with a consultant doesn't mean that that skill is passed on to other people, or reapplied to similar problems in the future. Knowledge is like a pebble dropped in a pond: the resulting ripples are strongest at the centre, and progressively weaker as they travel across the surface of the water (Figure 11.2).

For the knowledge that has been transferred to the individual to become useful to the organisation as a whole, it needs to be captured so that other people can make use of it (made available through the company's Intranet, for instance); it needs to be reused (perhaps by consciously deciding to move the newly-skilled individual to other parts of the organisation). But clients rarely plan with this in mind.

It's therefore not surprising that – in recent years – clients have been moving away from this style of consulting. The tangible impact of process consulting is hard to see, let alone quantify. In an environment where clients are finding themselves under increasing pressure from their shareholders to deliver performance improvements, it makes sense to reposition consultants as doers – people who can be made accountable for the value they add. It's rather as though clients

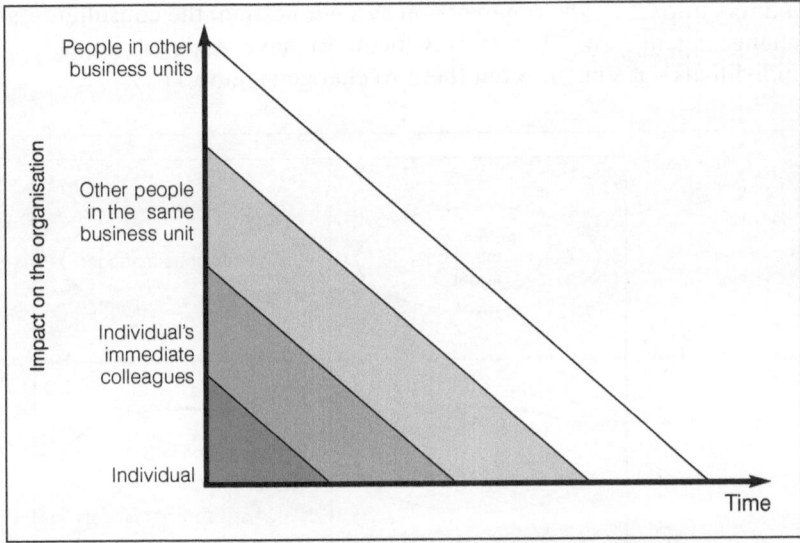

Figure 11.2 *The typical pattern of knowledge transfer in a consulting project*

have said: 'Enough! We now know that the intellectual capital we do acquire from consultants only benefits a small part of our organisation for a short time, so you might as well get on and do it anyway.' But the drawback here is that you're dependent on the consultants to keep the change going. Take away the consultants who are running the new systems, and the new systems cease to run. Drop the consulting firm that's developing the new product, and there won't be a new product. No doubt, therefore, there'll be a backlash to the backlash at some point in the future, and clients will, once again, talk about reducing their reliance on consultants by acquiring the latter's skills.

Rather unsatisfactorily, we've ended up in an environment in which the pendulum of fashion swings from one extreme to the other, and never seems to find a point of productive equilibrium at which to come to rest. The impact of the consultant-as-facilitator is too limited in scope; that of the consultant-as-manager is too temporary.

The solution was supposed to be change management, in which consultants could be employed to 'do' something specific but indirect – to manage change. Looking at the experience of many companies, you could argue that this approach has combined the worst of both worlds. If the consultant-as-facilitator's job is to have an indirect impact on individuals and the consultant-as-manager's job is to have a direct impact on the organisation as a whole, then the consultant-as-change agent's role has largely been to have a direct impact on individuals – it's been to tell them to change (Figure 11.3).

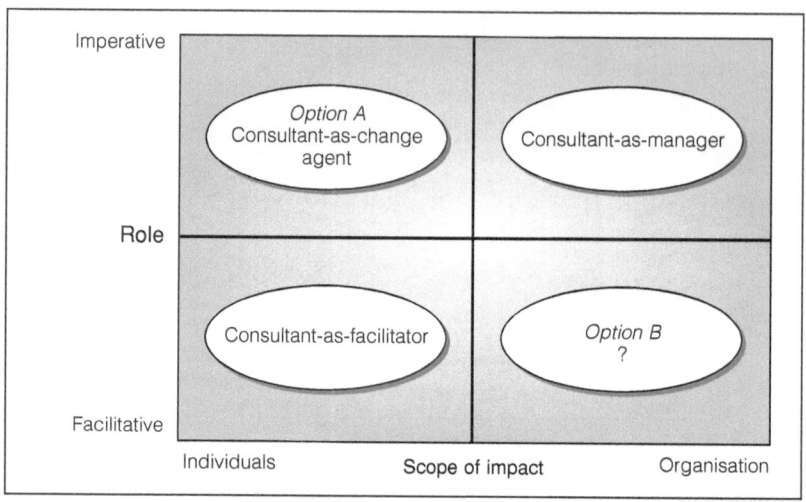

Figure 11.3 *The options for breaking out of the facilitation-manager extremes*

That's just as much of an oxymoron as telling someone to be empowered: the effect of telling them instantly disempowers them. It's likely that a solution that applied facilitation (to secure lasting change) to organisations as a whole (to ensure breadth of scope) would be much more effective. But what does this solution look like?

The Road to Damascus?

To make a large number of people do something, without telling them to do it, requires a very different form of consulting to those described above.

Essentially, it's consulting that engages with people emotionally, as well as intellectually, that makes them want to do the right thing.

The eminent business anthropologist, Geert Hoftstede, spent decades unpicking the different drivers of behaviour, eventually categorising them in terms of their depth of influence. Market influences, he concluded, were the most superficial; those from our families, inculcated since childhood, were by far the strongest.

The problem for consultants is that the whole history of consulting has been at the most superficial levels, as perceived by Hofstede – focusing on market or organisational drivers for change, based, fundamentally, on a rational model of business. Harking back to its roots in Taylorism and ideas of scientific management, consulting has been about extracting from particular experiences general rules that can be applied from business to business, and from industry to industry – processes, frameworks, benchmarks.

'Now, what I want is, Facts ... Facts alone are wanted in life', Mr Gradgrind famously demanded at the opening of Charles Dickens' *Hard Times*. For decades now we've assumed that facts are the only way of making an organisation change what it does. Yes, facilitation works, but only at an individual level: if you want an entire organisation to do something different, you have to present it with facts. Facts about new markets, about competitor threats, about efficiencies other companies have achieved. What are burning platforms, other than facts? Facts are like brick walls that we bang up against, forcing us to change direction: you can't argue with facts. Of course, 'changing direction' isn't the same as 'changing'. Just as you can be the same person rebounding off a wall (bruised, perhaps, but otherwise unchanged), the organisation that changes its strategy remains the same organisation. But the more significant the fact (the larger the wall, if you like), the more likely the fact is to change an organisation, as well as its strategy. If the burning platform is sufficiently severe changes make take place that result in a different organisation emerging from the ashes. Run into a large enough wall fast enough and you may do yourself lasting damage (although you may also learn that running into walls is not a sensible or sustainable course of action). A key part of the way in which consultants have sought to galvanise organisations is to present them with facts: the bigger the facts, the better.

But this isn't the only model. If we're looking for what plugs the gap in terms of effecting facilitative change on an entire organisation, we may need to look to very different sources of inspiration.

Religions have an excellent track record of changing people's lives, but they don't rely on facts. Quite the reverse: in the absence of proof, they rely on faith. If you look at what happens when a new chief executive takes the helm of an organisation on the brink of collapse, a pretty similar pattern of events emerge. Typically, new CEOs don't turn up with a detailed blueprint for change squeezed (in several volumes) into their briefcase. They're much more likely to have a high level vision of what should happen, what they want to achieve. And their first actions are more about performing miracles – making things happen that the organisation didn't believe it was capable of doing – instead of 'rational' management tasks. They tell stories – 'myths', 'parables' – which elucidate the meaning of these miracles and which articulate a set of values to which their people are asked to subscribe. These stories are then taken up and embellished by a core group of believers – and it is these believers, not the chief executive, that develop and own the detailed blueprint of the new organisation.

It's hard to think of a process that is more different to a conventional consulting project – indeed a conventional project of any sort. But could consultancy look and feel different? Why shouldn't it be as much about ritual and experience, as rational analysis? Yet probably the only way in which the behaviour of people in an organisation can be changed in the long-term is at a far more profound level than most consultancy attempts. It's not so much process engineering, as value engineering; it's not so much giving clients a serious piece of analysis which they can't ignore, as about giving them some sort of – literally, life-changing experience.

Of course, some individuals and some small firms are experimenting in ways of doing this. They might use drama or drawing, for example, as a means of providing a neutral environment (the play or the picture) in which management issues can be thrashed out obliquely, through role-playing rather than actual confrontation. But such ideas have yet to find – to my knowledge – any widespread take-up (which is why this chapter is the exception to all the others: there is no best practice here). Why? Part of the resistance is cultural: it doesn't look or sound particularly business-like to talk in these terms. In fact, recession is driving us further towards the fact-based view of consultancy because it's more quantifiable, clearer cut. Fact-based consulting also carries with it a (spurious?) illusion of predictability – and hence control – about it. If we show Company X the same report that we showed Company Y, Company X will behave in the same way that Company Y did. When things don't work out quite so smoothly (and they often don't), we blame the vagaries of fortune (a poor presentation, a difficult client), rather than abandon the quasi-scientific model we have come to rely on. And yet, the irony is that it is precisely this model that is currently depriving the industry of any substantial legacy. The consulting industry will only start to effect genuine change among its clients when it can change itself.

And that is the focus of the next part of the book.

Part 3
The Right Way

12

Organisational Design: Delivering Solutions, Not Services

There'll be a further broadening of what we mean by 'consultancy', caused by consulting firms adding new services to their existing portfolio, particularly in outsourcing. Why would a parcel distribution company, like FedEx, launch a consulting firm to launch a consulting service? Because it realises that it's delivering 'solutions' as much as parcels. No business in its right mind is today defining itself simply as a provider of a discrete service. You don't go to your travel agent simply to purchase a ticket, but to have someone who can provide a 'solution'.

Betsy Kovacs, Chief Executive,
Association of Management Consulting Firms

There seems to have been another revelation in the consulting industry recently, with lots of people leaping up shouting 'we deliver solutions, not services'. At first sight, it seems a pretty cosmetic change – after all, putting together customised services for clients is what has largely fuelled the growth of the consulting industry of the last 20–30 years – but the practical implications are rather more substantial. Creating and delivering solutions on a sustained basis, for entire markets rather than single clients, pre-supposes that your organisation can redirect its resources as opportunities arise. Rather than having to pull people out of different parts of a rigid organisational structure

What clients want:
To work with consulting firms that are able to reconfigure their resources to meet the clients' requirements

Value-based consulting:
Having an organisational structure in which flexibility is balanced with the need to exert control

What consultants want:
To have manageable organisations

137

into one of those unconvincing multi-disciplinary teams where no one has actually worked with anyone else before, the 'solutions'-based organisation treats its people as a portfolio of different skills, allowing them – amoeba-like – to reform where market needs dictate. 'Companies are increasingly unwilling to sacrifice size and breadth for market responsiveness or vice versa', argued a recent article in – appropriately enough – the *McKinsey Quarterly*:

> Companies are now organising to realise the benefits of both … Rather than viewing the corporation as a portfolio of business units, their managers regard it as a portfolio of resources and opportunities to create value. This opportunity-based design perspective gives these companies the flexibility to bring the most useful resources to bear on the most promising opportunities.[1]

It sounds perfect from the client point of view – organisations that effectively form and reform themselves on their behalf. But the reality of making it happen is far from perfect, at least from the consulting firm's perspective. Smaller firms have a distinct advantage: with less money and emotion invested in a formal structure, it's much easier for them to contemplate – and manage – a much more fluid organisational structure. But for large-scale firms, there are already so many problems to getting thousands of people moving in one direction that giving them *carte blanche* to move in whatever direction they thought clients want seems like a recipe for chaos. 'How can a firm with so many consultants seriously expect to be so fluid, and still retain any degree of structural coherence?', asked one senior partner in a large firm I spoke to. 'We do try to give our consultants as much autonomy and individual responsibility as we can', commented another, 'but we still have to have a clearly defined structure. Without it, we couldn't take decisions, we couldn't make any strategic investments; we probably wouldn't know what was going on.'

But perhaps there's a third way – a way of enabling a consulting organisation to create more and deeper internal links between different parts of the formal structure, rather than throwing away the structure altogether. To some extent, such links already happen, often as a result of an individual or business unit that has an idea that will have to involve other parts of the consulting organisation to be successful. Informal networks, voice and e-mail, chance meetings, can all be requisitioned to build an informal consensus around a specific opportunity. But it's hard for such activity to take place on anything other than the most micro of micro levels. Unsanctioned by the powers that be, the initiative will only receive recognition – and investment – once it's already taken off. In this way, many consulting firms operate an implicit, but strictly Darwinian process of natural selection: the 'fittest' projects make it through and are rewarded as a result. It's consultancy 'red in tooth and claw'. But, by the time the initiative has fought its way up the bloody evolutionary ladder, an opportunity to create critical mass in an emerging market may have been lost: evolution may be an effective way of stripping out the weak, but it's hardly fast. If it's to be more successful, the 'third way' has to recognise and manage these cross-business links even in their most fledgling form and 'incubate' them if it's to match the speed and flexibility of smaller, more fluid organisations.

Michael Goold and Andrew Campbell are two UK-based academics who've studied how businesses do and don't collaborate internally.[2] From their research, they've identified six different types of cross-business links, a taxonomy that proves illuminating when applied to the consulting industry.

- *Shared tangible resources:* The divisions of a manufacturing firm might share access to a single research and development department; the branches of a car dealership might share a single procurement team. The problem with applying the idea of holding tangible resources in common with the consulting industry is that the vast majority of those resources are peripheral to the core consulting process – unlike R&D in a manufacturing company or procurement for a car dealership. In a consulting company, it's the functions that consultants cannot do for themselves that are typically shared – IT support, bulk copying of documents – so the reason for their being shared is negative. Yes, there are economies of scale, but these are comparatively unimportant in a culture where the unit of most calculations is the individual. The idea of shared tangible resources is the equivalent of outsourcing the non-core consulting activities from the individual consultant to the corporation. It's a way of getting rid of a problem, not a means of maximising an opportunity. There are even some central functions within the consulting firm which, although they should be about creating shared value in theory, are tarnished with the same negative brush in practice. Training and development is a good illustration of this. Continuous professional development is something that goes to the heart of good consultancy: consultants don't stay still, they learn ahead of their clients so they can filter out 'best practice' and apply it. But personal development is also something that needs centralised support – how else would you know what training was available, whether it suited your needs, what the alternatives were? But, take it away from the consultants in total, and it rapidly becomes 'impersonal' development – something someone does to you, rather than you taking responsibility for doing it for yourself.

- *Shared know-how:* For a distributed organisation, it makes sense to have what used to be termed 'staff functions', central teams whose remit was to share best practice in specialist areas throughout the organisation, rather than expecting individual business units to develop such thinking in isolation and with fewer resources. But this is equally a model that works less well when applied to a consulting firm, as it comes up against two problems. The first issue is that there may not be much in the way of overlap when you look at the knowledge bases of different business units within a consulting firm: retail consultants need to know about retail; financial services consultants, about financial services; even within financial services, the knowledge required by an expert in retail banking is hugely different to that of a specialist in, say, the re-insurance market. Common knowledge which is also valuable knowledge is a contradiction in terms. But the second, perhaps more fundamental issue is that, where a knowledge overlap exists, that knowledge is polarised into one of two forms. First, there is generic business knowledge and consulting skills: it might be appropriate, for example, to have a team whose job it is to train people in standard accounting principles, or presentation skills. Like 'impersonal development', this type of knowledge typically enters into a vicious circle from which it's difficult, if not impossible, to emerge. Generic knowledge is perceived by the consultants to be less valuable than their specialist knowledge, and because it is less valuable it can be delivered by less valuable (that is, less expensive) people. Equally, because it is delivered by less valuable (less expensive) people, this knowledge begins to be seen as even less valuable. And so on, pretty much until everyone gives up in disgust. By contrast, the second type of knowledge, although it too may be generic, is seen to be very valuable to the consultants – too valuable, indeed, to allow non-consultants anywhere near it. This second type of knowledge is usually much more informal and unstructured than the first; typically, it refers to cultural rules and conventions – ways of doing things, rather than codified principles. Again, it becomes a means by which the individual business units within a consulting firm – and the individual sub-units within them – erect barriers between themselves and any possible connection with the rest of the organisation.

■ *Pooled negotiating power:* Given the size of some consulting firms, it's not surprising that this is one form of cross-business collaboration that does happen. Large consulting firms are some of the largest users of airline services in the world; they're also huge purchasers of IT equipment and services. There have also been attempts to extend this collaboration into their client base, by passing onto smaller clients, with less negotiating muscle, access to the bulk discounts a firm has been able to negotiate on its own behalf. As such, this provides a potential means by which consulting firms can build relationships with potentially fast-growing companies that could not otherwise afford the former's consulting fees. The problem here is that the actual level of collaboration this approach involves is comparatively slight, especially now that – for example – web-based procurement systems can replace the organisational interaction and cultural integration such activities would have required in the past. Technology may have increased integration behind the scenes but has had almost the opposite effect on the surface – giving individuals a greater sense of control and autonomy by being able to input information themselves, rather than having to route their request through a central purchasing department.

■ *Co-ordinated strategies:* Consulting firms – large and small – have always built teams from different parts of their business to meet specific customer needs, but one of the legacies of the e-business consultancies is that the higher the level of skills integration you can achieve, the more value you can create for clients. Consulting firms have therefore been able to utilise their client-focus to provide localised momentum, pulling people together with a common goal. This, then, is quite the mirror image of the barriers to collaborative working highlighted above: because this is client-related work, people are much more prepared to share thinking and lessons. The problem is that it's difficult to replicate this type of collaboration the further you move away from client work. Thus, co-ordinated strategies may – and do – work effectively at the level of individual projects, but don't usually work across consulting organisations as a whole. Client imperatives drive synergy – but they are also the reason why synergy does not take place more widely. Initiatives have to be attached to a known client if any synergy is to be created; 'unattached' initiatives have no independent impetus.

■ *Value-chain integration:* Like co-ordinated strategies, it's proved hard for consulting firms to do anything more than integrate a series of partners for specific opportunities (see Chapter 13). Once again, the further the reason for integration is divorced from an actual client need – improving efficiency of delivery would be a good example – the harder it is to achieve integration in practice.

■ *Creating new business ventures:* These are where companies use the synergies they have identified both within their business and with external partners to develop new markets, products and services. From the standpoint of consulting firms, there's a logical progression: from co-ordinated strategies (a low level of synergy developed to meet the needs of a specific client); through value chain integration (a medium level of synergy developed for a small number of clients); and finally to creating new business ventures (a high degree of synergy for many clients – that is, a market). You have to have a substantial client rationale in order to create a high level of synergy.

In summary, the approaches which consulting firms can adopt in terms of creating synergy can be mapped across two dimensions (Figure 12.1).

First, there is the level of integration involved – a venture might involve only a few people with different skill sets for short amounts of time, or it might stretch to people from many different disciplines working together over long periods. Second, there is the driver for integration – whether this is internal, for a single client only, or for many clients at once. Sharing tangible resources and know-how, and pooling negotiating power, do not have a sufficient impetus from client work to generate anything other than low levels of integration. Integrating value chains and developing co-ordinated strategies work well, but tend to be limited to opportunities with one or a small number of clients. The greatest levels of integration can be achieved by aggregating the needs of individual clients into a discernible market.

So the key to the 'third way' – building consulting firms that have the flexibility of niche players to create solutions, combined with the

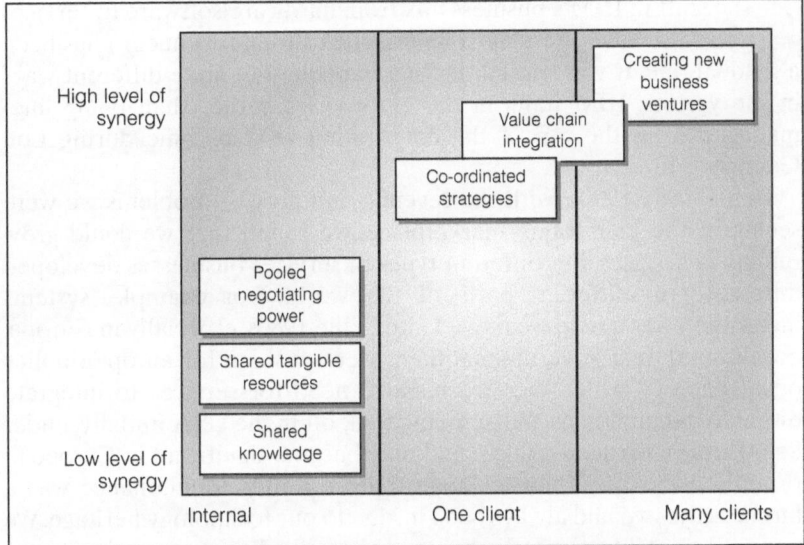

Figure 12.1 *Correlating levels of synergy achieved with client focus*

size and scope of the largest firms – lies in aggregation of demand, being able to add up the needs of individual clients into visible, addressable markets.

Delivering Solutions: From the Very Large ...

One of the companies people are most likely to mention, when they're talking about solutions-based organisations, is IBM. 'When Lou Gerstner took over IBM in 1993', argues Mike Howarth, Worldwide Director of Marketing, Business Innovation Services, at IBM Global Services, 'he pushed the company into providing solutions for business problems in a formal sense. Almost a decade after his arrival, we've gone from being a very small player in this arena to the world leader through a lot of hard work and clear determination. Today we have over 150,000 services professionals. We got to where we are by maintaining a strong commitment to putting the customer first, everyday. We let them do most of the talking. We listen, provide our point of view, and then execute. I think only a handful of companies are going to be able to deliver the kind of value that global customers need.

❛ The shift in IBM's business mix from hardware/software to services (today, services are almost half of IBM's total revenues)) has been a gradual one. It was something that happened in quite different ways in the various IBM units in the early years, rather than being line-managed from the top of the corporation as it became during Lou Gerstner's time.

In the early 1990s, with the revenue and growth problems we were seeing in the mainframe marketplace, we found that we could grow rapidly in services, but different types of services businesses developed differently in different parts of the world. For example, systems integration was a field we moved into in the 1980s, especially in Europe, because national governments there were pushing for an open policy on technology which drove a massive need for services to integrate disparate technologies. Outsourcing took off in the USA initially, under Sam Palmisano's leadership – and only became significant in Europe in the mid 1990s and then in Japan quite recently. Maintenance was a business that we had always been in, due to our technology heritage. We had traditionally operated a bureau processing business migrated over time into a networking services business. In 1992, we made the conscious decision to move into business consulting on a worldwide basis.

IBM Global Services was formed as a distinct group in 1997 to pull together all of these different services units around the world, and the decision was made to organisationally report it directly into the CEO as opposed to reporting through the sales organisation which had been the previous structure. In the mind of the software and hardware sales people, once the equipment or software had been sold, that was the end – they didn't see themselves as accountable for the long-term results. Today, we have a different sales model, one that recognises that consultants tend to have first or second call on a senior business executive, and that more people need to be involved if we're to deliver a solution, rather than just complete a sale. We're evolving a more consultative client management model, not a pure sales rep structure which was our heritage.

When we launched the IBM Consulting Group – in 1992 – it was a stand-alone entity. Systems integration was distinct from consulting and was viewed as more 'blue collar' work, more part of the IBM fabric, but we finally – in January 2000 – merged these two services, because that's what clients had been asking for – integrated services, speed to value.

Organisational structure is one of the key dilemmas for all professional service firms. Should our approach to market be co-ordinated by industry, by geography, or by business function? Consultants tend to see things in industry verticals; systems integrators, in terms of horizontal functions. The real answer, of course, is that our structure has to reflect the client's view, not ours – it's clients, after all, who know what they want and see what work's been done.

Moreover, the Internet and e-business have challenged many of our preconceptions about the boundaries between industries and business functions and the IT systems which can improve or transform them. We constantly have to strive to add more value to our clients' businesses. What puts us in a unique situation is the considerable heritage on which we have to build that value. That gives us, we believe, an enormous advantage over other firms – the big five and some of the strategy consultancies – some of which are only just now trying to major in the technology area. Historically, we have always been strong on the technology side of the equation. We've put in a tremendous amount of effort into bolstering the business knowledge side of the equation – and we continue to do so. In fact, earlier this year we were the only public services-led company to rank in the Top Ten performing companies of Business Week's Info Tech 100 listing, due to 'big demand for consultants who demystify and make the most out of tech gear'. We've been publishing an increasing number of 'white papers' on strategic business issues to highlight our business expertise. We've also acquired a strategy consulting firm – Mainspring – which has been integrated into our Strategy & Change consulting practice. But no one ever questions our ability to do systems integration – and that's a heritage that no other consulting firm is going to be able to match. Our consulting value proposition is built on expertise at the boundary of business and IT – where each discipline interacts with the other in a mutually reinforcing way. To deliver that requires a combination of star people with business and IT expertise, plus intellectual capital that underpins it and technical ability to implement it.

Organisational change in IBM has been primarily driven by customer needs. Those customer needs can change almost overnight. In fact, that behaviour shifted again recently, with the economic downturn: clients are much more interested in short-term efficiency propositions, rather than market expansion ideas. This downturn, combined with shifts like Y2K, the dot.com boom and crash and e-marketplaces, have made the last 24–30 months the most turbulent that we have seen in the last 25 years in this business.

In the early 1990s, consultants were novelty items, interesting people you could drag along to sales calls. In the high growth days of the mid to late 90s, consultants had relative independence and freedom to find clients – their margins were high and subsidised by IBM. With the recent downturn and the shift in value propositions we have led, there's been a shift: consultants aren't outside the overall value proposition, they are an absolutely integral part of it.

Unlike five or ten years ago, if you joined IBM Global Services now, you'd see all the base management methods of professional services now in place. The consulting practice effectively started off as a greenfield site: we had to engineer – not re-engineer – all of the core parts of professional services automation, and we had to standardise them across the world. Then the emphasis has moved towards linking these to IBM's existing system – its opportunity and sales management systems, for example. And, for the last couple of years, the focus has shifted again – towards changing those existing systems to reflect that we weren't dealing with product transactions but business management engagements.

Now, we're in a situation where IBM has made many of the internal changes necessary to make it a *professional* services firm, and it's other consultancies that are trying to replicate our model. Bottom line, what we see in the marketplace these days is that customers want implementable solutions that are tailored for their precise needs, not just independent advice. IBM's not the slightest bit shy any more about pushing our own technology when it's clearly in the best interest of the customer. Nor do we back away from using a competitor's technology if it's a better fit. We used to go to great lengths to prove that the IBM consulting group was independent and agnostic. Now, there's no need, because what customers crave more than absolutely independent advice is an ability to deliver speed to value. Consulting is part of IBM's core, not an afterthought.

What were the factors that enabled IBM to make this change? I'd say it was the fact that the solutions-driven culture grew from the bottom up, but was matched by top-down commitment. We've seen a real step change at the top of this organisation. For much of the 1990s, IBM was under pressure to create packaged applications, to compete head-on with companies like SAP. That was a business we decided to get out of in 1999, and it was an extremely important decision from the client point of view. Clients prefer a customised solution to packaged things. To have tried and tested skills, but to be able to customise what you do for the needs of individual clients, has been something that put us in the position of being a trusted business advisor. **'**

... To the Very Small

The Virtual Development Group was founded in 1995 by Laurence Udell in the belief that organisations need to be flexible, to operate in a radically new way. The company is an organisational and management consultancy, which specialises in strategic thinking and implementation; crucially, it is itself a practical example of the kind of virtual organisation it helps its clients develop. The company is run by a core team of around ten people; work is delivered to the client using members of the 'core team', 'extended team' members of whom there may be 15–20 at any one time, and a wider group of 150+ Extended Partners whose skills may be utilised less frequently.

According to Martin Clemmey, one of the company's Directors, 'virtual working isn't just about remote location working and new technology – it's primarily about the attitudes, behaviours and relationships needed to work across cultural, departmental, company and geographic boundaries. A virtual perspective enables each solution to be delivered by a unique and temporary team evolved exclusively for that client or project.

❛ We're in an era of transition. There's no real consensus about what the paradigms of the knowledge age will be: we're still using 'old economy' language to describe something that's only just emerging. We hear lots of glib statements that we're living in an idea-driven, not a sequential economy; that we need to be able to manage paradox and uncertainty; that change is the only constant. It's a strange mixture of post-relativism and pre-industrial ideas of co-operation. And we see that in practice. We're working with organisations that are still largely based around hierarchical business units but which are trying to respond to rapidly changing market conditions. Faced with this, we can't just jump into the top of hierarchy, when someone's spent 20 years getting there, and claim that everything's going to be different in the future: we'll be shown the door. What we have to do is to meet them on their own ground, to find out what they know and don't know about their business, and then get a debate going. We don't go in saying we're a virtual company, or holding up a big placard saying we're different. It's just something that will come out in the course of conversation: they'll notice that the person they're dealing with doesn't have a job title. Organised as we are, we try to hold up a mirror to people.

The Virtual Development Group is structured around a nucleus of core members, who assess the requirements of each project individually and recruit from within a global network of virtual members, who are themselves practising business leaders, consultants, tutors and academics, with a proven track record in their field. This structure enables us to tap into a dynamic and very experienced skill set which would not normally be available within other consultancies. It's an approach that is designed to ensure that every element of the solution is customer-focused and results-orientated.

We believe that there are three significant benefits to clients from virtual working – speed, flexibility and quality. Speed comes from the fact that small teams of experts – like ours – are more capable of working in tandem, but more independently than teams of bright, but less experienced consultants. Quite simply, this means that we can do more things simultaneously: we don't have to wait for one task to be completed before we start on another. Flexibility and quality come from the way in which we build teams. Bigger consultancies usually have teams dedicated to a particular sector, but we can pull in highly-qualified individuals and create bespoke teams of experts. Getting the right talent engaged in this way brings a greater richness to the solutions created.

At all stages of the consulting process – from initial sales call to handover – it doesn't matter if someone is fully employed by the company or a member of our extended team. Obviously, in the early stages of building a client relationship, we'll use a lead consultant who's done a lot of work with us before, but the other people on the team are selected on the basis of the match between their skills and the work in question. Co-ordination, communication and knowledge sharing are all crucial. Every project has a project co-ordinator, who's effectively the relationship manager, responsible for managing the project. This is important, not just because we're a distributed group of people whose input has to be co-ordinated, but also because one of our differentiators is speed of delivery. We often find the CEO saying that he's been to a conventional strategy company that says the project will take nine months, whereas he actually needs it in as many weeks. The project co-ordinator then goes to the resource manager with a list of the skills required, and the resource manager searches the database for availability and come back with a short-list from which we build the team.

From the consultant's point of view, working in this kind of environment can be disconcerting. We have to bring in new blood all the time, and people have to hit the ground running. It's hard to generalise about projects, you might have a situation where a professor of strategy from a major European business school has to work with a professor of marketing from a different school and a team of e-commerce specialists. We'd expect people to meet together at the beginning and then occasionally during its course, so they have to be able to develop good working relationships with the other team members very quickly. Virtual working also makes demands on the individual in a number of ways through flexibility, tolerance of pressure, management of time, space and conflict. This means that members of the virtual team must manage themselves, use their authority and take accountability when appropriate based on their expertise and capabilities, irrespective of perceived status, and not rely on the capability of a designated leader to lead.

To help in this process, we've developed own in-house collaboration tool – there was nothing on the market that met our very specific needs – which is part knowledge management, part education, part document management, including facilities for detailed project management. And we reinforce this with a training programme, helping people to work in virtual teams.

But the real 'glue' that binds us together is our values. As part of the process of signing up to a project, every team member has to identify what personal learning they'll get from it. We believe that the best client work is done when the aims of both client and consultant are aligned, when both will gain from the project being a success. One of our key values, therefore, is that everyone has to be able to learn from a project. Another is that everyone has to commit personally to a project: they have to be completely accountable for delivering their individual part of it. This has to be an ongoing commitment: virtual teams can't function effectively, if all the members wait to do their work at the very last minute.

Everything is branded as the Virtual Development Group, as that embodies our values. People – clients and consultants – want to work with us because we undertake and deliver demanding, challenging and ground breaking solutions both in what we deliver and how we deliver it. At the moment we're expanding at a rate of 60 per cent a year. The model we've evolved can cope with turnover doubling in a single month – the tap can be turned on and off very quickly. **,**

Organisational Design: Towards a Value-Based Approach

Creating a solutions-based organisation requires a combination of external pressure from clients and internal top-down prescription.

Where pressure from clients is not matched by any coherent drive for change from within, consulting organisations are likely to find themselves being buffeted by successive management fads, responding piecemeal, often to the wrong priorities. Where there's insufficient pressure from clients, consultancies are going to find it difficult to create and sustain the impetus required to integrate the different parts of their firms. One without the other won't work.

---

[1] Russell Eisenstat, Nathaniel Foote, Jay Galbraith and Danny Miller, 'Beyond the Business Unit', *McKinsey Quarterly*, 1, 2001.

[2] Michael Goold and Andrew Campbell, 'Desperately Seeking Synergy', Harvard Business Review, 1998.

13

Value Chain Integration: Building Constellations

The challenge going forward is the same for consultants as it is for their clients – how to combine the best of the 'old' and the 'new'.
David Yates, General Manager, American Management Systems, Europe

Value chain integration is one of the ways in which consulting firms have been able to reconcile what they perceived to be conflicting demands from clients. It's one of those magic terms that suggest that someone somewhere is adding value. If you're lucky, it's your organisation, but it doesn't really matter too much as long as someone in the chain is doing it. For the small, specialised firm, collaborating with different partners means that you have access to projects which would have been beyond your individual capabil-

What clients want:
To have access to world-class skills

What consultants want:
To be able to access new markets and to manage sudden spikes in demand by building relationships with software vendors and specialist firms

Value-based consulting:
Clarifying the role that the consulting firm plays, and by integrating the skills of different partners around specific propositions

ities. For larger, more broadly-based firms, it's a way of getting access to the highly-specialised skills your client wants without having to acquire or develop them yourself.

Technology vendors have therefore looked for higher-end consulting firms who can give them superior access to potential clients, and consulting firms have sought out companies whose technology is likely to dominate future markets. But one of the lessons of recent history is that this is a strategy that's not quite so simple in practice. The term 'value chain' implies a set structure, with

151

each organisation forming a comparatively distinct, definable role within the chain itself. But, in a volatile environment, strategy

> is no longer a matter of positioning a fixed set of activities along a value chain. Increasingly, successful companies do not just add value, they *reinvent* it. Their focus of strategic analysis is not the company or even the industry but the *value-creating system* itself, within which different economic actors – suppliers, business partners, allies and customers – work together to *co-produce* value. Their strategic task is the *reconfiguration* of roles and relationships among this constellation of actors in order to mobilise the creation of value in new forms and by new players. And their underlying strategic goal is to create an ever-improving fit between competencies and customers.[1]

The 1998–2000 e-business boom reinforces this point: the original 'pecking order' of the consulting industry was inverted: technology vendors, who were accustomed to having much lower level contacts within organisations than management consulting firms, suddenly found that the doors of much more senior executives were open to them. Changes in technology – and the innovative way in which some companies were exploiting these changes – challenged the classic sequence of strategic implementation – that you formulate a strategy, build the operational capability to deliver it and then develop the systems to support your operations. Suddenly, technology was something you had to think about in tandem with strategy, because it offered opportunities and raised threats which would have been inconceivable a short time earlier. That the dust of this upheaval has by no means settled is demonstrated by the continuing debate, within the consulting industry, about the relationship between technology vendors and consultancies. Should consulting firms work to ensure a seamless transition from strategy to implementation by building close links with technology firms? Or should consulting firms remain objective, able to proffer disinterested advice? Some would claim the two approaches are not incompatible – that it is possible for a consulting firm to have close links with the suppliers of technology while continuing to be independent; others would vehemently deny this, claiming that any alliance compromises a firm's objectivity.

Moreover, the issue has been fudged by consulting firms that set up multiple alliances – another of those have-your-cake-and-eat-it strategies – defended on the basis that the individual alliances minimise the gap between strategy and implementation, and the collective multitude of alliances ensures that a firm is not too beholden to any one partner. The multiple alliance model has fudged, too, the difficulties that have perennially beset the consulting industry, of making partnerships work. It's never been easy to align the conflicting goals of those involved: consulting firms typically want to provide the implementation support for large systems work; technology vendors may have their own consulting teams; and so on. Fighting skirmishes over these issues on many fronts means that you never really have to engage in a sustained battle to solve the problems with just one company. It's not so much value chain integration we've seen in recent years, as value chain spaghetti.

As we untangle this spaghetti, it's likely that consulting firms will shift to one of four more distinctive approaches to value integration, depending on the level of integrated working and the number of partners involved (Figure 13.1):

- *'Splendid isolation' firms* have no formal links with any other company; the value they bring to clients is their guaranteed objectivity.

- *'Broker' firms* will also have no level of integration links with other companies, but will have an extensive network of contacts they are capable of marshalling for a particular client project; their role, however, will be almost wholly one of co-ordination – putting the client in touch with expert firms rather than managing the process of integrating the latter's input.

- *'Single issue' firms* will have highly integrated working relationships with just one or two other companies, extending to joint product development and planning, as well as sales, marketing and delivery.

- *'Orchestrator' firms* will be less integrated with their partners than 'single issue' firms – typically, collaboration will be limited to sales, marketing and delivery – but they will maintain these relationships with far more partners; effectively the prime contractor in client relationships, the role of 'orchestrator' firms will be to integrate the work of different partners, so that the project delivery appears seamless from the client's point of view.

Each strategy will have its own advantages and disadvantages: absolute objectivity may mean that a firm is well-positioned to help clients in the earliest stages of their decision-making processes, but is much less likely to be involved in the detailed implementation – a

Value-Based Consulting

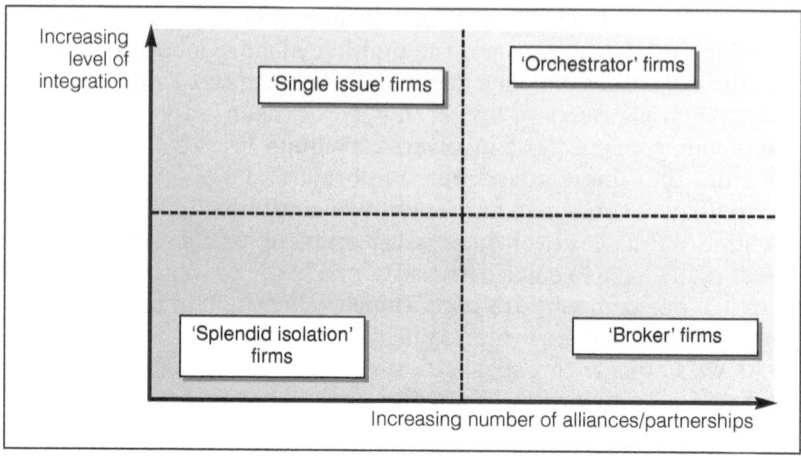

Figure 13.1 *The four emerging strategies for creating value 'constellations'*

position that most strategy consulting firms have already adopted. The strength of 'broker' firms will lie in their ability to know which specialist firm to deploy in which area, whereas that of the 'orchestrators' will be in integration.

From a Jack of All Partnerships, to a Master of One

Erik Sandersen, is responsible for global partnerships at IconMedialab. 'These days', he says, 'everyone seems to want to be a partner with everyone else. They think it's important to have as long a list of alliances as possible, but you have to wonder: what value do they really get from them?' IconMedialab found itself asking exactly this question in early 2001.

❛ By the end of 2000', Sandersen says, 'we'd amassed more than 30 global alliances, many of which didn't mean much beyond the local level where they'd been established Someone in Paris might decide they needed an alliance with a particular software vendor for a particular project: overnight we'd have a global partnership which other people weren't interested in. So we decided we wanted a much smaller number of alliances, but ones that actually had some value. It sounds obvious, but it actually involved a lot of soul searching. Consulting firms like to be agnostic, they like to be able to do everything for everybody, but we had to accept that we were only 1,500 people and that we couldn't be good at everything. At the same time, we didn't simply want to become the implementation arm of a vendor, and we didn't want to be a value-added reseller. Although we would be the smaller partner, we wanted a complementary relationship. Crucially, we wanted to be able to go to market together.

Of the alliances we had at the time, one stood out – with IBM. IBM was an attractive partner for several reasons. First, we were familiar with their products in many of our offices, so the level of retraining we'd need to do would not be great. Second, as the software market consolidates, it seemed to us clear that IBM's product range was gaining ground and likely to be central on an ongoing basis. The third reason was the breadth of IBM's product and service range. There was – and is – some overlap between what they do and what we do, but, by and large, IBM is such a big organisation that we found it perfectly possible to work with the software divisions without treading on any toes on the services side. One of the first questions people asked, when we announced our new partnership with IBM [in July 2001] was 'why not Hewlett-Packard?' HP doesn't have the same scale of services business, so in theory, it would have made for a more synergistic match. But the potential conflict between ourselves and IBM's Global Services Division hasn't been significant in practice and, in any case, was outweighed by a fourth and – from our perspective – most important reason for allying with IBM, which was IBM's approach to partnerships. We'd already found that many companies make nice noises about partnerships but relatively few of them ever translate words into actions. IBM not only had people whose job it was to develop partnerships – and make them work – but also had a much more two-way attitude, and that was something we found to be unique among the bigger players. And why should IBM be interested? What was distinctive about IconMedialab? The answer's a simple one: in the hey-day of e-business, they'd signed up hundreds of web-integrators, but we offered a far more comprehensive European coverage in this sector than any of their other partners.

Once both sides had agreed to the partnership in principle, the next step was to work out how it could be made to work in practice. We each had dedicated people working on this, and – having learned the lessons from our previous partnerships – we made sure that there were people, again from both sides, in every country who were responsible for working out what the alliance would mean for them, developing revenue targets, working out how we would track and handle sales leads. It helped enormously that IBM has a separate partnership organisation whose job it is to open doors within the rest of IBM's otherwise intimidatingly large and complex organisation. But opening doors would only get us so far. Why should someone buried deep in IBM's organisation worry about getting business for us? They may not even know we exist. The onus was therefore on us to get to them.

We realised that what we needed was a 'product' that we could sell internally in IBM, and use this as the means of establishing the value we could add. After much research and internal debate, we came up with the concept of the corporate communications platform. Essentially, this combines Icon's skills in design and technology integration with IBM's products: the idea was to make the software sales people at IBM realise that Icon could help them make their revenue targets, but to develop something that would bring business in for both them and us. Icon came up with the initial idea, but we then tested it out with people at IBM to see whether it made sense to them both in terms of the market and commercially. The companies then jointly developed the idea, so that the original blueprint could be tailored to the way that IBM is accustomed to working. If we hadn't done that – brought people together to create a combined product – then, not only are we likely to have encountered a degree of 'not invented here' resistance, but it might also have been difficult to find the right people to involve and motivate from IBM's side. We had to have answers to the kinds of questions in which a software sales person is going to be interested; in a sense, we had to be able to talk IBM's language. From then on, it's been a case of marketing the product internally within IBM.

In some ways, it's hard to imagine two more different companies: a large US-based company with a complex hierarchy, and a European one which is small enough to be much less formal, with much less in the way of structure. But we haven't found either the cultural or geographic differences a problem. In fact, I think each side has something to learn from the other. IBM is more bureaucratic and perhaps could use some of our flexibility, but, frankly, Icon has a lot to learn from IBM in terms of how it can manage its operations efficiently. Having a link with IBM is also helpful to us when we go out to talk to clients – they're less concerned about our financial strength – and we can inject a little creativity into IBM's image. What matter most, I think, is that we've managed to establish very effective relationships at a personal level, with individuals in different countries already working well together.

Of course, with the partnership only a few months old, it's too early to know whether we've succeeded or not, but we believe we've found a sustainable model for the future. There's a lot of collaboration going on at a local level, but it's hard to aggregate that up into global results at the moment. One thing we believe has helped is our willingness to be clear about what we can and cannot do. From IBM's point of view, the conventional consulting line of 'we can do anything' is immensely obstructive. This means that the key challenge we face is remaining focused. In their hearts most consulting firms want to be generalists, and we've had a lot of internal debate about the implications of what this alliance means to us. But, if the consulting partner can do everything, how do you know where the real points of synergy are? **)**

Keeping the Link with Spin-Out Organisations

The concept of value chain integration hasn't just been complicated by a more volatile environment in which understanding your fixed position in a stable hierarchy matters less than your contribution to a dynamic 'constellation' of companies. Value chain integration – even value constellation integration – has largely been thought of in terms of pulling separate companies together, but, since the late 1990s, many large corporations have used the different look and feel of Internet-based companies as a rationale for side-stepping the more intractable cultural, political and bureaucratic problems of their main organisations by creating spin-out companies to grow new lines of business. Here, the challenges of integration are rather different: too little integration, and any synergy between the parent and spin-out will be lost; too much integration, and you risk eroding the desired differences between the two.

It's as much an issue for consulting firms as it is for their clients. Many established firms chose to set up separately branded organisations, aimed at delivering more creative services to clients under a more 'new economy' image.

'No one said that working across conventional boundaries was going to be easy', says Paul Nannetti ruefully. As head of CRM and DareStep in Europe, he was responsible for building Cap Gemini Ernst & Young's Darestep operation in Europe, and was previously responsible for e-business in the UK organisation. Nannetti has spent the last three years working out how best to capitalise on opportunities created by Internet-related technologies.

' DareStep in Europe was set up in 2000, following the successful launch of the organisation in the US, although there were some lessons we'd learned from the American experience which meant that we wanted to do things slightly differently. After all, unlike the US, we weren't starting with a greenfield operation: prior to the merger with Ernst & Young in 2000, Cap Gemini had already set up a separate e-business consulting division and in 1995 had acquired Bit-IC, a web design and technology business, in the Netherlands. DareStep was set up as a design and usability division, with a strict focus on helping clients define and achieve an optimal user experience for their applications, both external (web, idtv and so on) and subsequently internal (Intranet, other internal applications such as ERP systems).

A key aim of DareStep was to bridge the gap between strategy and technology: it was about the art of the possible. 'My strategy is this', people were saying, 'but help me define it visually, in order to better express what we're trying to achieve, and to understand the technology implications. How can I make the technology attractive to people, so they'll actually use it?' Our creative work needs to be very practical and grounded. One of our clients was a major retailer which wanted to build a network of kiosks offering a broad range of additional services based in their stores. This wasn't something you could think through in abstract aesthetic terms: you had to take into account that the customer might be computer illiterate, so you also had to consider how intuitive the system would need to be, for an inexperienced or once-off user. We combined our usability, design and ergonomic skills to build a touch-screen interface that, when tested with pilot groups in our usability lab, was highly successful in that even people with no computer experience felt comfortable navigating through the services and found the experience 'fun'.

Moving from an initial focus on e-business or web related projects, we quickly realised that the issues we were dealing with were far broader: 'usability' is as much an issue with an ERP system, as it is with a web-site. Just like a web-site, the success of an ERP implementation ultimately relies on getting people to use it.

We took the decision to establish DareStep initially as a separate division, conceptually and physically. We wanted it to develop its own brand, under which it could define its differentiated capability to clients and recruits, recognising this would enhance the brand of CGE&Y itself. Initially, we based DareStep's UK operation in one of CGE&Y's old offices in a rather unfashionable part of London, but we quickly realised that this wasn't going to work, if we were to recruit the creative talent we wanted. So we relocated to the heart of London's media district and created a much more relaxed and informal office space. But, at the same time, we also needed to ensure that DareStep didn't become too remote from the parent company: we didn't want to create a Frankenstein that would go off by itself. So we had to learn to work together, to collaborate.

So, although this new office space was initially for DareStep employees only, we gradually moved strategists and technologists into the adjoining areas, with the aim of building informal connections between these sets of people, something which, in turn, would foster mutual respect and a willingness to share learning. Even though DareStep was part of CGE&Y, we still encountered challenges in working together. Consultants are pretty results-focused people. They feel they have to achieve something new every day, and they've developed processes to help them do this as efficiently and quickly as possible. Creative people are very different, they tend to potter around a lot, looking for that inspirational moment as it were; asking them what they've achieved today is something they're not so used to – one breakthrough idea for them is valued more than a whole series of drafts or proposals. We also found that our languages were very different: I remember one meeting which had gone on for four hours before we realised that we were talking at cross-purposes because we each understood something different by key words like 'template'. Distinctions like this were difficult to deal with, but they're something we've persevered at. In time, frustration has given way to mutual respect.

One of the things that's kept us going has been top level management support. We established a strong management team for each local DareStep operation, led by a CGE&Y person, not an outsider. We had to have someone who understood and had empathy with the parent company, who was capable of pulling DareStep into the rest of the organisation. Below this layer, the creative people we hired almost all came from the outside: we had to build up credibility internally and externally, so we recruited highly experienced people from media, advertising and technology companies. But we also went out of our way to find people who were mature and established within these professions because we felt they were more likely to make the transition into a mainstream consulting environment. At the micro-level, we identified a core of people who would work most closely with DareStep: we didn't try and build collaboration across the board.

But perhaps the single, most important thing we did was win our first piece of work. That gave us enormous confidence – and internal credibility. In the first months that followed, almost all the projects done were stand-alone Darestep work: they didn't really need to work with anyone from elsewhere in CGE&Y, but we knew, even then, that the market wasn't going to be in stand-alone solutions of this type. Much more important was the market that linked this kind of work to larger-scale integration with legacy systems, and organisational and process redesign. But people have gradually appreciated that the work we do may open doors to much more significant consulting projects. Looking at how customers interface with a business is such an important subject that you tend to develop very deep, long-term relationships with your clients.

We've had to overcome many challenges – the increased commercial risks, for example. We had people (colleagues and clients) questioning why building a site with DareStep costs more'. And we'd have to say: 'well, building a site that's interesting and intuitive to use requires a lot more effort than just coding a basic screen … but you get the payback by increased and more effective usage'. It's very difficult to defend that position until you've shown that you can deliver something that's worth the wait. There was a lack of empathy between the consultants in CGE&Y and the creative people we brought into Darestep: the former tended to relegate the latter to backroom functions – the kind of thing that would make our creative people simply down tools and leave. Overcoming this has required strong creative leadership, which in itself takes time to build, and a lot of internal selling.

We've now got between 400 and 500 people in Darestep, but we still learning how to make it all work. With the initial learning process behind us, we have the confidence to integrate the DareStep organisation much more tightly into our core structures. Increasingly, we are tending to focus the Darestep specialists on the earliest stages of a project – its incubation. Once we get beyond this, we reduce the size of the Darestep team, and pull in the technical and operational skills required from other parts of CGE&Y's business. Essentially, we've become better at integrating the new and more traditional competencies. And that model – an integrated technical and creative solution that's consultancy-led – is hugely attractive to clients: they get the innovative thinking, the specialist know-how making it happen, but all subject to the kind of rigorous controls and project management disciplines they'd expect from a consulting project.

The challenge now is to industrialise this collaboration: we can do it, but not as quickly or efficiently as we think we should be able to. **"**

Value Chain Integration: Towards a Value-Based Approach

Multiple alliances have diminishing returns, and the onus is now on consulting firms to develop much clearer strategies in which their role is better defined, and in which the way the alliances add value to clients can be more clearly demonstrated. Some firms will chose to be completely independent; a minority will have the scale and scope to become 'brokers' and 'orchestrators'; but, for most firms, the challenge will be to move from the plethora of alliances they now have to focus on one or two issues, with one or two partners – they have to become 'single issue' firms. And 'single issue' will mean precisely that: concentrating on one marketable idea around which there can be joint product development, as well joint marketing. The value to clients lies in the former; the value to consultants lies in the latter (Figure 13.2).

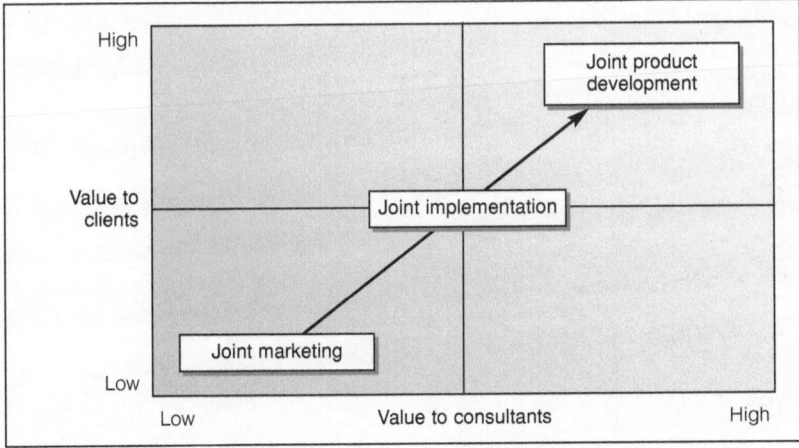

Figure 13.2 *Using alliances to add value to both consultants and clients*

Much of the value chain integration that has taken place within the consulting industry to date has been driven by the needs of consulting firms themselves. Alliances with software vendors have ensured that the consulting firms are in a position to help clients with implementation; alliances with specialist consulting firms have provided alternative reservoirs of skilled labour when demand exceeds supply. The real value to clients will come when partners with different backgrounds, different skill sets collaborate to develop and apply new services and products.

This is equally true where collaboration happens within a firm, across different business units and different disciplinary boundaries. One of the strategic thrusts of the late 1990s was to spin off groups of people into distinct companies: the challenge for the early part of this century will be to reintegrate those skills without losing their specialist perspectives, and to be able to collaborate, as Paul Nannetti puts it, on an industrial scale.

---

[1] Richard Normann and Rafael Ramirez, 'From Value Chain to Value Constellation: Designing Interactive Strategy', *Harvard Business Review*, July–August 1993.

14

Technology: Bridging the Gap from Plan to Practice

It's been a truism of management that technology should not determine business strategy: technology should be an enabler, never a driver.

But it's equally true that any period of significant technological change – such as we've seen in relation to e-business – creates openings for companies whose competitive advantage lies in their ability to exploit leading-edge technology ahead of their rivals. For these companies, technology may – quite legitimately – drive their strategy. But the advantage is rarely more than a temporary one: most of these companies find it hard to maintain their innovative edge as the technology environment matures and are overtaken by rivals with other, more sustainable sources of competitive advantage.

> **What clients want:**
> To have a seamless transition from business strategy to technology implementation

> **Value-based consulting:**
> Finding new ways to resource projects while maintaining an integrated service from the client's perspective

> **What consultants want:**
> To provide a seamless service which neither cannibalises its fee rates in strategic work, nor beomes prohibitively expensive when it comes to implementation

This is an issue, too, for consulting firms working in this area: uncertainty about the role of technology in recent years has opened the door for primarily technology-based consulting firms to reposition themselves to do more non-technology related consultancy.

'Clients that are looking to realise value from their IT systems will turn more and more to technology-based companies, rather than to management consultancies', opined one senior consultant – working, not surprisingly, for a technology-based company. 'As a technology company, we've been effectively running our own e-business for years', commented an executive at another technology company that has now launched its own consultancy, 'and we therefore understand the technology from a practical perspective. That generates a lot of trust among our clients: when we talk about whether a particular technical architecture will or won't work, they know we're doing so in the light of this experience. You can contrast this with the hype they read about in the press: we help them focus on the practical issues.'

But – if the past is anything to go by – the ability of technology-based firms to move upstream, towards strategy work, will only last while the uncertainty about the role of technology remains. It's probable that, once technology has settled back into its historical position as an enabler, rather than a driver, that technology-based consulting firms will return to their historical – supporting – roles. Moreover, we'd also expect to see the operational-based consultancies entering the technology firms' own space, by taking on more specialist technology work as the maturing market makes it easier for the former to invest in specific skill sets, as happened with both **BPR** and **ERP** consulting. The conventional development cycle for the technology consulting industry therefore has three phases (Figure 14.1):

- *Phase 1* – in which uncertainty over the role of technology paves the way for technology firms to offer generic business advice;

- *Phase 2* – in which maturing technology ceases to be a strategic driver, enabling the strategic and operational consultancies to re-establish their presence in their core markets of generic business and technology advice; and

- *Phase 3* – in which further technological maturation enables these broader based firms to compete with the technology firms on the latters' 'home' territory.

The transition through these three stages can be justified in terms of offering a smoother transition from the high level business and technical strategy to its detailed implementation. Moving from theory to practice has always been an uneven process in the consulting industry, with some projects either being abandoned or substantially changed as it becomes apparent that the initial recommendations will be hard, if not impossible, to implement. But this is a problem that has been particularly acute when it comes to IT consulting. Here, the divisions between paper and practice seem much more significant, and

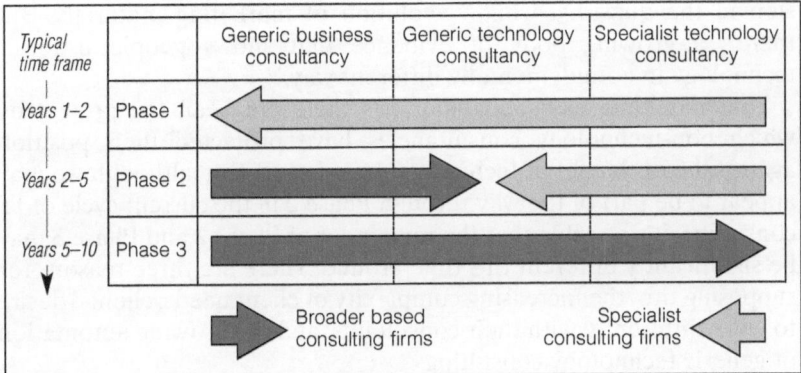

Figure 14.1 *The three traditional stages of evolution in technology consulting*

the skills and working practices of the suppliers required to deliver them seem more diverse. From the consulting point of view, these three evolutionary stages have had the added benefit of allowing firms with high charge-out rates to take over markets at a point where the market is actually starting to mature. This is in contrast to most markets, where maturity produces a decline in the unit price.

In terms of e-business, the fight back by the strategic and operational consultancies has been underway now for a couple of years. But it's not being fought on the superiority of strategy over technology, but on a different premise entirely, that it is now difficult to divorce a business proposition – whether it is business-to-consumer, business-to-business or business-to-employee – from the medium in which it is communicated. The user (customer, supplier or employee) experience is central to realising the technological promise.

'One of the main problems in e-business', said one executive I spoke to, 'was that people tended to start from the technology and work backwards into the business model. What they should have been doing was analysing how their business model was going to make the market work in a different way, how participants – suppliers, partners, customers – will behave differently. This isn't easy because it's hard to visualise: you have to look at what these stakeholders actually need and then assess the extent to which the assets of your company and emerging technologies could meet these needs. It tends to be much easier to imagine how you could use a particular piece of technology – so this is what most companies do.' 'It's tempting', agreed another interviewee, 'to see web-based technology in particular as just another step in the quasi-Darwinian evolution of marketing materials. But there's a growing body of evidence that shows people use this technology in a fundamentally different way.'

Understanding user behaviour has therefore been the pretext by which non-technology consultancies have protected their position against the onslaught of technology-based ones. But, although we may appear to be part of the way through Phase 2 in the current cycle of IT consulting, it's possible that the remainder of Phase 2 and Phase 3 may be significantly different this time around. There are three reasons for supposing this: the increasing complexity of client needs; clients' desire to work differently with their consultants; and the growing automation of generic technology consulting.

'Companies', said one consultant I talked to, 'tend to see technology in linear terms: they still think of it as a chain: $a + b + c + d$. They define success in terms of how well the technology they have chosen performs at just the first stage, on the assumption that, if it performs well there, then it will continue to perform as you add more partners and more functionality. Very few companies can define at the outset the total numbers of participants in a network.' Another highly sought-after feature is modularity. 'There's a huge amount of functionality on offer', commented a senior executive. 'The market's changing so fast, both in terms of the business we're trying to create and the technology enabling it, that it's hard to believe that a single supplier will be able to supply all your needs in the future.' But both scalability and modularity may have to be sacrificed in the name of integration. According to one specialist in financial services: 'typically a corporation has had a good idea and has developed a technology strategy around the benefits of a particular piece of software, without giving much thought to the integration issues. What this means is that – however good the original idea – organisations end up replicating physical processes in the digital realm, exacerbating existing problems rather than resolving them. The continued departmentalisation of many large corporations makes the situation even worse: we've certainly come across companies that – ironically – have ended up further away than ever from being able to collate disparate sets of information into a single customer "view".'

Complexity, integration and the successful balance of the two requires more sensitive project management than has perhaps been the case with IT projects in the past. One consultant, discussing the factors determining the success of projects in this environment, commented that project designers had to be 'sensitive to the politics of the organisation and evolve a relatively distributed project structure into which people were co-opted for given periods of time, instead of setting up a large, central group of people whose work becomes inevitably alienated from the mainstream organisation'. Another agreed: 'there was a false premise that integration revolved solely around the software. In practice, the actual software was – and is – only one small component of a very complex picture which involves existing systems, future business objectives and corporate culture.'

The technology consulting market at present differs from those of the past in that it is likely to become more complex over time. This has significant implications for consulting firms: as long as the technology market refuses to coalesce around a small number of leading products, consulting firms will find it difficult to prioritise their investment and resources, and many will be reluctant to over-commit themselves to a specific vendor in the fear that they may make the wrong choice, allying themselves to one of the many potential losers in this area, rather than one of the few winners. It also makes it harder for them to enter the generic technology consulting market by their traditional means – technology strategy. While issues like scalability and modularity add up to a greater need for integration at a strategic level, it's likely to be bottom-up integration, rather than top down (which is the approach traditionally – and inevitably – taken by non-technology consultancies).

We need to add to this the changing expectations of clients, who want technology (and technology consulting) to deliver results more quickly than in the past. 'You have to be able to break projects down into manageable components', stressed one IT director, 'partly so that you can test them out in practice instead of describing them in theory, but partly also so that you can change your approach rapidly, as circumstances evolve'. Clients are therefore looking for organisations that can demonstrate concrete evidence of their track record in particular environments and packages. 'We've found that clients simply don't listen to you, if you go in talking about theory or the functionality of particular systems', says one technology consultant. Experience is key'. While this attitude on the client side obviously stacks the deck in favour of the technology specialists – the consultancy divisions of software developers, for example – it doesn't preclude the broader-based consulting firms altogether. By forging alliances with what they believe will be the winners in the technology race, the latter believe that they will be able to demonstrate that they have access to the skills clients demand. The real problem for non-technology consultancies lies in the fact that the quick delivery times and concrete track records clients want are both predicated on a different *modus operandi* – something that may prove a far more difficult obstacle for these firms to surmount.

As one client put it, 'the business model of the very large consulting firms really doesn't work when it comes to the technology aspects of e-business ventures. It's still far too focused around long-term projects and carries with it a significant management overhead. You really need a much more entrepreneurial approach.'

The sour aftertaste of Y2K and years of ERP implementation persists, paving the way for technical specialists who promise more efficient, more focused delivery. The criticisms weren't confined to the broader-based incumbent consultancies: many of the new entrants of the last few years, in stressing the customer-centric characteristics of much web-based technology, turned out to be long on theory and short on practical detail. 'We looked at a whole range of firms', commented one client, 'and found that there wasn't much difference between the new start-ups and the "old economy" firms. This seemed to result partly from the fact that, in order to meet demand, the former were recruiting staff from the latter, and partly from the way in which the newer firms were seeking to replicate the processes of the bigger firms, presumably to give themselves more credibility with clients who expect that kind of thing. We also find it hard to understand what any of these firms would actually do: there's a lot of talking around the subject, but no one seemed able – or prepared – to tell us the three main problems we'd encounter and what we could do about them.'

Part of the dissatisfaction with the role of generic consulting firms stems from the fact that 'generic technology consulting' – the main bridge built by such companies to give them access to the specialist IT consulting markets – is being automated. The process of encroachment began with business process re-engineering – something that it's now possible to do using a software package rather than by hiring a consulting team – and has gradually led from there to the mapping-out supporting technology. Some of the mainstays of generic technology consulting – package selection, for example – can now be done through a judicious combination of in-house resources and online exchanges and decision-support software.

In the past, the emergence of a new, disruptive technology has enabled technology services firms to adopt a broader, more strategic role with their clients. This role, however, has been short-lived: as the new technology becomes assimilated, the specialist firms have found themselves returning to their traditional, more limited role and the more general firms have acquired the skills needed to support the technology. The first stage of this cycle has been true for web-enabled technologies, with technology consultancies finding clients more willing to hire them to undertake a wider range of work. But the second stage, in which generalist firms occupy the specialist ground, is more doubtful: first, because of the inherent complexity of e-business technology needs; second, the operational models of many of the larger firms do not lend themselves – in clients' eyes – to working in this more fragmented environment. Instead, that market for technology consultancy which has historically served as a bridge, allowing specialist technology firms to offer generic business consultancy, and management consultancies to enter specialist technology markets, will increasingly become a barrier in both directions (Figure 14.2).

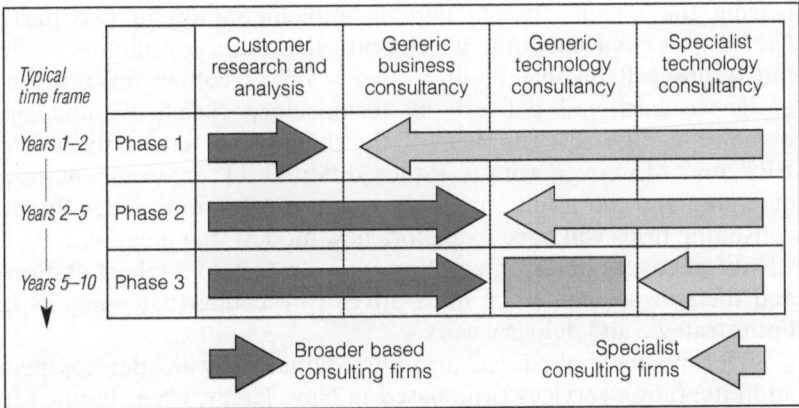

Figure 14.2 *The three stages of evolution in business technology consulting*

This suggests that the emerging technology consulting market, rather than becoming a huge monolithic market – in the way that ERP technology was – may remain the domain of smaller, specialist firms. The importance of being able to demonstrate a proven track record in implementation will also be a major barrier to entry for the larger firms. Smaller, specialist firms will also thrive better in this environment: culturally, they'll have the more entrepreneurial, hands-on feel that clients are looking for; financially, they'll be more able to handle a stream of small-scale projects than larger firms looking to deploy junior staff for long periods of time. Nor does this mean that there will be no role for the larger, more generic consultancies within the e-business technology market. A fragmented market will create its own – albeit different – opportunities. In particular, a fragmented market will not be able to direct or integrate itself, and there will, therefore, be a significant market for technology programme management and integration services – a market that many mainstream firms will see as being part of their core competence.

Combined Onshore/Offshore Development

While that bridge, between what is essentially the front and back ends of an IT implementation, may have been a very valuable one from the perspective of some parts of the consulting model, it was never satisfactory from the client's point of view. Its structure wasn't strong, and it frequently collapsed under a weight of unfulfilled expectations, leaving the systems design and its implementation in two parts. Creating an environment in which both clients and consultants win in the future will involve finding other ways of connecting up-front design to back-end delivery by technical specialists without the monolithic, high-cost supplier relationships of the late, unlamented ERP era. Moreover, with a rather different IT environment now emerging and the traditional three stages of evolution under threat, consulting firms will have to explore new models that meet both their clients' needs (seamless integration from strategy to implementation) and their own commercial imperatives (protecting their margins in both strategic and delivery work).

Silverline Technologies is an international software development and integration services firm, based in New Jersey, USA. Formed in 1992, it now employs around 2,400 software professionals worldwide, with more than half of them located in a number of delivery centres in India. Dan Hayter is Silverline's Executive Vice President of Sales and Marketing for Europe; Chris Baker is the company's European Managing Director.

❛ ***Chris Baker:*** We work with organisations at all stages of an IT project, from the early thinking and design – where we can bring consulting expertise and help them through their technology choices – into the building phase where the applications are delivered. Often that is a combination of implementing packaged software and integrating the new software into the existing systems. Finally, there's the back-end maintenance, which is the upgrading, management, and running of the completed systems. Silverline's model allows it to bring services right through that lifecycle. To be fair, the epicentre of all of this is in the application development building phase. However, we also have consultants for the front-end design work as well as capabilities at the back-end that allow us to maintain applications for clients.

We operate in a number of vertical industry sectors, but financial services is the most significant, with clients like Fidelity Investments, First Data and JP Morgan. We're also active in a number of other sectors, such as manufacturing (where we work with Volkswagen of America, General Motors, Ford Motor Company and ALSTOM), fast moving consumer goods (FMCG), retail and pharmaceutical and healthcare. Our clients tend to be the large global operators, and our relationships tend to be global in nature: we often start working in part of the clients' business and grow into other business areas and geographies through a combination of delivering good services and good account management practices. By doing so, we build a good reputation internally as a valued service provider.

The model by which we deliver those services to customers around the world has three levels. Some of the work is done at the client's premises, largely using local people – project managers, business consultants and technical architects. As a client moves into the IT development phase, we tend to move the architectural and design work into our local regional development centres such as the ones in the UK. From there, the bulk of the work can be moved into India, but only when a client is comfortable with this strategy.

Taking IT development work offshore is not new to many large companies. In fact, many of them have tried it, even if only experimentally over the last few years, and have seen mixed results. The lessons from this are that organisations were wrong if they thought that they could simply bundle up a piece of work, sub-contract it to a company in India and just leave them to deliver without too much ongoing co-ordination. Naturally enough, given the rapid style of IT development these days, the results were often pretty disappointing: there was a disconnect between what the client actually wanted to do (and how that changed and moved along the way), and the way the system actually delivered.

Dan Hayter: I think it's worth saying that India nearly always delivered very fine code in double-quick time. The problem was that, because the contract had effectively been fired off and forgotten about, people felt that all the normal checks and balances you have in IT projects (done in the West by Western companies) weren't needed. The assumption was that India was some kind of magic place where you sent work and that it got done, and came back perfect.

People started to realise it wasn't that simple – and that's what's behind our operating model. We've taken the wonderful Indian offshore model and added on top a great low cost, quick, highly IT-literate environment with the normal Western quality controls such as project offices, project management, and an understanding of customer requirements and business issues. We think this is the only way you can effectively manage a large client with a particular business problem, that is itself made up of a multitude of other minor business and technical problems, and bring everything together in order to deliver a solution on time.

Many of us here have backgrounds in the big consulting or IT firms. From those experiences, we've been in environments where there is the capability of making sure projects are done to the highest quality and to the highest standards, on time. We've worked with people who have understood the business issues. But what that environment always lacked was the engine room to get the work done in a cost-effective manner. Now we've put quality standards and a low-cost offshore model together to have the best of both worlds.

Chris Baker: You can split the market into three. First, there are shrinking numbers of organisations that haven't used India or an offshore supplier yet. Show them the facilities, the level of training of people, and the quality levels out there, and it's relatively easy to wow their socks off. There are then those who have tried offshore development in the past and haven't been happy with the results. Our position is the one that Dan's just mentioned: that we've not only got the first class engine, we've also got people who understand you, your market place, speak your language, plus all the necessary co-ordination between. The third segment is those people who are already using offshore development and are very happy with it. Here again, we think we've something to add, because we have both the consulting front-end and the low cost delivery engine from India. Essentially, we're positioned between offshore firms that can already deliver quite satisfactorily out of India but don't have the business front-end expertise, and those that have a fantastic business front-end but have nothing in India. We've some quite long mission statements floating around, but I like to boil it down to saying we do more for less.

Are there benefits for clients beyond cost savings? The quality of programming is undoubtedly high in India. More than a decade ago, the country's whole educational system was turned on its head with the aim of producing the highest possible standard of software engineers. One of the differences between India and the West is that, every few months, programmers in the US or Europe will start thinking they want to do something new or different. This means that people move around too rapidly to allow them to specialise. And, because skills were scarce, it was quite possible to get promoted even while your level of expertise was low. In India, it's very different. Technical training has been gone about in a far more business-like consistent way. People's skills are much more formalised and they tend to do things by the book. The Indian government and organisations, like our own, that train lots of people in IT skills (not just own employees), are also aware that we have to conform to all of the various international standards so you get consistent quality. Again, that's a big difference to the West. Finally, you've got just the sheer numbers. In Europe and the US, you often struggle to find certain skills; in India you've got access to a number of software professionals who are qualified and have a strong work ethic. Many tend to be eager to do a great job, rather than just the next job.

Dan Hayter: Does that mean that India is the only place that can provide such resources? We're going through an exercise at the moment to evaluate other possible bases. One is in the Asia Pacific region – the Philippines in particular; the other is Central Europe. But, in practice, there are relatively few countries where an operation like ours would work. Several factors have to be in place – a large, educated workforce, a safe and reliable infrastructure, access to training resources and a basic ability to do business. India has the advantage of being a mature player in this field and a high level of English. It's going to be hard for relative newcomers, like the Philippines, to match what India offers right now.

Bulgaria, for example, has some fantastic IT people. They're very smart. They're the ex-Soviet Union's IT domain. Each country in the Soviet area got something to do – and Bulgaria's job was replicating Western IT technology. They went on to become pretty good at most IT and were largely responsible for the Soviet space programme.

So there are some genuine rocket scientists you can employ in Bulgaria! They've got the quality, just not the quantity. But it's not a large country, and the percentage of people with these skills is very small, so you very quickly run out of them.

Chris Baker: In addition to the issues associated with hiring software professionals, we have to pick our clients carefully as well. The word 'partnership' is often bandied about at such moments, but, in an offshore environment, you need a very tight-knit community of people dealing with a project to make it work. People from the client side, local people from our side and then the people in the offshore teams all need to work very closely together. Wherever the political environment inside a client is still one of 'me, client; you, consultant', it's unlikely to work very well.

Dan Hayter: It splits into two camps: there are people who understand how you use offshore facilities. They're usually the same people who've understood how you manage large IT projects. Once they understand that, it's possible to incorporate an offshore component into the work process. But there are some very big companies who still see offshore facilities as an opportunity to chuck a problem over the fence to a third-world country, where they can get it done cheaply – and that's where it invariably goes wrong. We have walked away from people who aren't willing to take on their own shoulders some understanding of what needs to be done and responsibility for doing it – and we will do so again.

There are some clients who, at the outset, say that they don't think offshore is going to work for them. But, if they understand that buying professional services is more than a transaction and that they're entering into a complex relationship, then we often find it's possible to move work offshore successfully, over time. For example, we're working with a big retailer who was initially completely against offshore work, and, in fact, the work we're doing with them hasn't got offshore yet. It was a big jump for them to take work out of their office and into an office that we set up next door for them. Once they were happy with that, we moved the work 40 miles down the road and now they're looking at moving this work to India because they can see how the model works. I think if we'd seen in that client earlier on that they were only interested in buying from us, we probably wouldn't have done the work at all. But, even from the beginning, they were talking the language of partnering, of being flexible and working together.

This kind of incremental approach helps everybody feel comfortable with each other and to understand exactly how the process is going to work as well as how we're going to run the project.

As the relationship matures, more work can move offshore. Probably the highest ratio of offshore to onshore work we'd look to achieve would be 80:20 or 85:15. We'd never expect a project to end up being totally offshore because then we'd be concerned that we'd lose the element of control clients require. The traditional offshore model was just: get it all offshore, worry about it later. But our business model is different from the traditional model. By having a combination of local coverage by consultants, we can offer them the programme management and control, and the high-quality advice they need, combined with a low-cost delivery engine. That's a model which people are finding very attractive – especially when faced with a period of protracted economic uncertainty.

At the moment, we don't see the consulting and high-end business advice market growing much beyond its current size, but we do think there's a huge growth opportunity in the later stages of the project lifecycle – from maintenance to business process outsourcing. Indeed, we're moving into that area already to take advantage of the fact that people are saying 'we're getting our IT done offshore but we're still employing lots of people and spending a fortune on back-office processes – why don't we do these offshore as well?' In most respects, these processes are even more manual and basic than the IT that they've already outsourced.

In our experience, when broader based consultancies go into IT, lots of things can go wrong. Up to now, they've been charging for what is essentially advice: now they have to deliver as well. Going from talking-the-talk to walking it is a big transition for an organisation and not all of them are making a particularly great job of doing it. As they've moved out of the advice role, a lot of the halo that they tend to have has started to crumble. The consulting firms face a wider range of competition that in many cases is better equipped to offer value to potential clients. At that point such consultants are extremely exposed, especially given their high prices.

There's an inevitable merging here of the technological choices to be made and the business choices to be made. It's no longer the case that you can figure out a business strategy and then work out what your technology strategy is. They're not separate any more, but one and the same. We don't really want to come in as just the artisans who will put the bricks together when someone else has chosen which bricks to use. We've seen that on too many occasions when a consulting firm has charged a lot of money and the resultant technology architecture is junk. So we like to be guiding the client from the start on both technology and business issues. We're not pretending that we're business strategy consultants in the sense of helping companies decide which markets to be in. But we are a notch just below: once an organisation has decided on a strategy, it still has to work out its business and IT processes. They're all one. **'**

Technology: Towards a Value-Based Approach

Clients want a smooth transition from business strategy to technology implementation, but they don't want to pay the same consulting rates all through the process. Consulting firms, on the other hand, want to protect their margins, and doing both types of work is one way of doing so. For firms in the strategic/operational segments of the industry, it's a means of charging more for technology skills than a specialist competitor might; for specialist technology firms, it's a means of earning higher fee rates than they might have to do in their core markets alone.

But, with a more complex and fragmented technology environment now emerging, none of these approaches seem likely to be sustainable. The priority, instead, will have to be to identify alternative models which are 'end-to-end' from the client's point of view, but which utilise different resource pools with different cost bases behind the scenes. It means – as Silverline effectively does – maintaining all the appearances of the traditional model of IT consulting, while finding radically different ways to deliver it.

15

Knowledge Management

[The strategic consultancies] are investing very heavily in ....
knowledge management and sharing – which mean that these firms
will be able to field the type of in-depth, cross-function and cross-
industry knowledge for which clients are prepared to pay premium
fee rates. The competition here will be all about finding the right
business model – one that enables the firm to invest and reap the
rewards of that investment very effectively.

David Pecaut, iFormation Group

What are the world's most common complaints about knowledge management? 'Where are the benefits?' has to be one of them, along with 'we've invested millions of dollars in knowledge management, but we still don't seem to have cracked how we leverage the really valuable knowledge – the stuff that goes on in people's heads'. Knowledge management – five years ago the panacea for every management ill – hasn't yet lived up to its considerable promise. For all the literature on the theory of knowledge management, there's been very little written on its successful implementation. We understand more than we did before about the obstacles to effective knowledge sharing, but we're apparently no nearer to coming up with a solution.

What clients want:
To be able to access the collective intellectual capital of consulting firms, as well as that of individual consultants

Value-based consulting:
Adopting a one-to-one approach, identifying a client's needs in terms of formal and informal knowledge

What consultants want:
To reduce the firm's reliance on individual consultants

Yet knowledge management is fundamental to the survival of consulting firms. It always has been, although in very different forms to 'knowledge management' as it is termed today. After all, what was the mentoring relationship between a partner and a small group of junior consultants if not knowledge management? The only difference now is that we expect much more knowledge to be˙ captured more systematically, and disseminated more widely, at least partly through the use of computers and the Internet. Moreover, in an environment in which clients' demand for specialist skills is making them increasingly turn to specialist consultants, the breadth and depth of the larger firms' intellectual capital has become one of the key alternative ways in which they can demonstrate value to their clients.

The situation is frustrating from a client's point of view as well. Why, when consulting firms are supposedly so rich in knowledge, does it take time for them to pull it together? Why don't specialists in different parts of the organisation talk to each other?

In 2001, McKinsey & Company published a survey highlighting what it found to be the characteristics of effective knowledge management and correlating this with superior financial and operational performance.[1] Managing knowledge management, it's argued, differs from managing other assets for six reasons:

- It's *subjective* – heavily dependent on an individual's background and the context in which it is used. Does everyone in an organisation have a common understanding? Do the same words mean different things in different places? Are there shared frameworks with which everyone can identify? High levels of subjectivity may be the result of cultural tribalism – a refusal to admit that a group of people is part of a larger whole – or, more positively, may stem from high levels of specialist technical knowledge being required in some areas. Whatever the source, organisations with high levels of subjectivity have significant internal barriers that have to be overcome. Not only does knowledge have to be distributed effectively and efficiently, but it also has to be useful (and it may be too specialised to be of value to anyone but its originating business unit).

- Knowledge can be *transferable* – it can be applied in different contexts, allowing many parts of an organisation to learn from the lessons of one. In many organisations, however, knowledge is applied in the business unit in which it is produced. Part of the reason for this is positive: a specialist business unit produces increasingly specialised knowledge, and this may be of little direct relevance to people in other parts of the business. Yet even the most specialist of specialist teams needs to be alert to the possibility that some of their knowledge may be applicable elsewhere. Facilitating the exchange of knowledge across functional boundaries is one of the most valuable roles an organisation can perform.

- Many organisations have a high level of *embedded* knowledge and therefore have to develop a means of balancing the structured, explicit information they capture with the unstructured, tacit information that typically resides in people's heads. They need to have the processes and technology, not just to make codified information available, but to put people in touch with experts who can share their uncodifiable experience.

- Effective knowledge management cannot take place in isolation: it has to be part of a *self-reinforcing* environment – one that publicly celebrates people who've made a substantial contribution to a company's intellectual capital and that penalises people whose behaviour does not promote knowledge sharing.

- Some types of intellectual capital are more *perishable* than others. A drilling company may have extensive data on how drills behave under certain conditions – information that's as unchanging as the rock structures through which the drills pass. Such data is also cumulative: because the raw material doesn't change, acquiring new information (for example, how a new drill bit behaves) doesn't mean you have to trash what you've already got. By contrast, other parts of the company's intellectual capital – its knowledge about its client (the contact details of who buys what, for instance) is only valuable in the very short term. It needs to be updated as soon as someone moves, if it's to be of any real use.

- It's tempting to treat intellectual capital as a stable commodity – reports and presentations that can be circulated. In reality, knowledge is not automatically a self-renewing asset: knowledge management processes and technology can stifle innovation, so it's important to inject a degree of *spontaneity* into the proceedings, making sure – for instance – that creative thinking techniques are used side-by-side with more conventional ways of capturing knowledge.

It's possible to see these six characteristics as the extreme ends of a continuum that highlights different forms of intellectual capital –

data, information and knowledge. Some organisations are more subjective than others. Some have greater levels of embedded knowledge than others. Some have a culture which reinforces behaviour, others don't. Every organisation has its unique profile: the key to effective intellectual capital management – as opposed to 'knowledge' management – is to design processes and systems that match this profile. Thus, an organisation with a high degree of embedded knowledge will need processes and systems that reflect this. By contrast, organisations where embedded knowledge is low (where, perhaps for health and safety reasons, procedures are described in minute detail) will need something different.

If we were to plot consulting firms on this continuum, consulting firms would score highly in all six of these categories. In practice, however, few of the strategies adopting for managing knowledge in consulting firms recognise any, let alone all, of the resulting issues. Instead, what we have is a situation in which the *process* used is de-valuing the *content*, so that knowledge is reduced to information, and in some cases data.

A combination of working in different specialist areas and an enduring legacy of a 'knowledge is power' culture mean that many consulting organisations lack much in the way of shared understanding – to use McKinsey's parlance, they're highly *subjective* and have a high level of *embedded* knowledge. For niche players and newer firms, the problem is less severe – focus and the fact of having grown up together mean that their people tend to have more common ground. They also have a smaller organisation through which knowledge has to be disseminated. For larger firms, grappling with subjectivity has been a much more serious problem. 'We've many different business units, all with their own particular focus, all seeing things from their own perspectives. Getting someone – say, a marketing specialist – to see the relevance and value in something a strategist or a technologist has said, seems almost impossible sometimes', commented one senior partner I talked to.

Transferability then, has been the single most important driving force behind most knowledge strategies – the need to prevent all those wheels being reinvented time and time again, by ensuring that everyone has been kept up-to-date with what is happening elsewhere in a firm. But it's something that has been achieved only at a considerable cost. 'It's Catch-22', said one partner. 'The more we try and make material accessible to all parts of the firm, the less specialised it becomes, making it less valuable to its immediate audience. The strategy has had the effect of reducing everything to the lowest possible common denominator.' Transferability has also, largely, been achieved at the cost of spontaneity: 'knowledge management has really become a way of recording what we've done, not a means by which we can become more creative', said one interviewee. Self-reinforcement, too, is low, as this interview continues: 'we did try to make a big thing of celebrating when people have had a good idea, publicising the idea, giving them scope to develop it further. But it's hard to keep this going: everyone thought it was exciting at first, but gradually they started thinking "oh, yeah, there's another idea – so what?"' The push for transferability at all cost, has been driven by concerns over the perishability of consulting knowledge, something which, in its turn, has been given greater momentum by the experience of e-business, when many of the incumbent firms were slow to respond to the new entrant e-consultancies. Clearly, some parts of consulting knowledge are very perishable – knowing the organisational structure of a specific client, for example, is only valuable as long as no one changes job; to be reliable, this knowledge has to be constantly updated. But other types of knowledge are much less perishable: a particular framework, properly developed, may be applicable to many organisations over a long period of time. We only have to look at the Boston matrix, or the idea of the learning curve, to see this.

Obviously, we have to be careful about making too wide a generalisation. Firms' reliance on codified knowledge – such as standard methodologies for delivery – vary widely in the different segments of the consulting industry. But, at the same time, this 'average' picture effectively defines what it is to be a consulting firm. An organisation with lower levels of specialist and embedded knowledge, and spontaneity, might be a provider of generic business advice – a government agency, say, responding to comparatively simple queries with standard answers. An organisation with much more perishable knowledge might be something more like a market research agency.

Summarising this (see Figure 15.1), we might conclude that the conventional emphasis to date, in terms of managing consulting knowledge, has been on transferability and perishability. All too often, the approach adopted to achieve these ends, combined with little creativity injected into the inputs to the process, and little sustainable attempt to reinforce the importance of knowledge management, has inadvertently meant that specialist, embedded knowledge has been reduced to the level of simple information, or even pure data (the expert know-how captured in a report or a proposal for a client, for example).

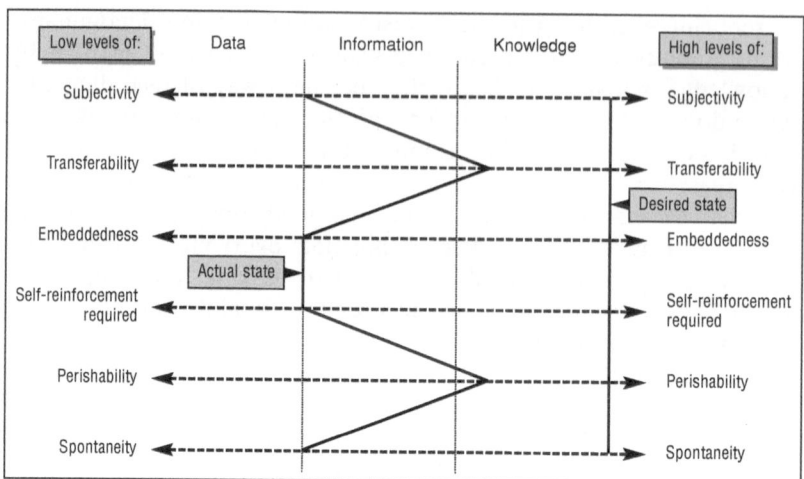

Figure 15.1 *Dream* versus *reality in terms of managing consulting knowledge*

Essentially, many of the difficulties with managing consulting knowledge stem from trying to bring knowledge to people, rather than take people to where the knowledge is.

Changing the Emphasis

This problem is compounded by the attitude of many consultants themselves, whose frustration comes from the fact that they perceive such characteristics as problems to be fixed. 'We should make our organisation less subjective by bringing everybody together', they say. 'We should reduce our reliance on embedded knowledge by codifying more of our intellectual capital.' Taken together, such imperatives mean that consultancies cease to be consultancies, and become something else: they're issues that go the heart of what a consulting firm is. In demanding improved efficiency and effectiveness in terms of knowledge management in a consulting firm, there's no recognition that, if achieved, a consulting firm may become something else. Reducing the interfaces in a firm may mean that knowledge can be shared more widely, but only at the cost of reducing the levels of deep, specialist knowledge in some parts of the firm.

Such an attitude is also counter to what clients want. While consultancies often see efficiency as the main goal, clients are looking to firms' knowledge management systems to provide improved access to the intellectual capital they require. In other words, they're looking for greater subjectivity, more embedded knowledge, not less.

Leif Edvinsson is the Chief Executive of Intellectual Capital Sweden and former Director of Intellectual Capital at Skandia, and one of the world's leading experts in intellectual capital. Formed by Edvinsson in 1997, Intellectual Capital Sweden has largely pioneered the idea of applying a standard methodology for measuring the intellectual capital of organisations – the equivalent of a Standard & Poor's rating for the knowledge economy. This methodology categorises intellectual capital under four main headings: the 'business recipe' is an organisation's business idea and strategy; organisational structural capital; human capital; and relational capital. Each category is represented in terms of its efficiency; the efforts made to renew and develop the intellectual capital, and the risk of deterioration.

Our research, says Edvinson, 'shows that there's a positive correlation between revenue growth and investment in structural capital. Knowledge databases and processes that capture and disseminate knowledge all mean that companies become less dependent on the contribution of certain, very valuable employees, and they're less exposed to the risk of those employees leaving. The impact of this flows through to the top line: companies that invest in structural intellectual capital are more likely to enjoy sustained growth. There's also an impact on the bottom line, as structural capital is a source of many efficiencies. Human capital can't work 24 hours a day, seven days a week: structural capital can.

As the ratio of human capital to structural capital in consulting firms is obviously very high, it follows that their results will be more volatile over time. But there's a twist here: the higher yield per head of companies that rely more on structural capital is achieved through greater efficiency – they're essentially focusing on getting people to work faster. With structural capital playing a less important role in consulting than in the other sectors we have researched, it's harder for consulting firms to expand efficiently. What structural capital doesn't give you is innovation. In fact, human capital is the only source of renewal in an organisation's intellectual capital.

The key, therefore, will be to break out of this cycle by using measures which better reflect the input of individuals in terms of areas such as creative thinking: in other words, it's to get people to realise that efficiency may not be the only goal of effective management of your intellectual capital. We could, for example, measure value-added per employee and use this as a benchmark. We don't, at the moment, because most accounting systems are not designed to report this kind of information. **'**

Towards a Value-Based Approach

So how can we reconcile the apparently opposing demands of consulting firms with their clients?

A starting point has to be for both consulting firms and their clients to be clear about their position with a framework like that outlined above, in Figure 15.1. There are some projects – and therefore some firms – for which efficient transfer of codified, formal knowledge is supremely important, such as IT implementation projects or anything where the consultant has to follow a prescribed set of rules. And there are many others where unwritten experience may be paramount, perhaps where difficult, strategic options have to be evaluated, discussed and agreed. And, clearly, there are many projects that fall somewhere between the two extremes, that are multiple shades of grey involving some formal and some informal components. The problem here is not really one of conflicting aims, so much as confused aims: neither side – consultant or client – has identified precisely what formal and informal knowledge is required where.

As a result, consultants find themselves trying to be efficient, where clients are really looking for effectiveness – distributing broad sets of information very widely, rather than providing in-depth specialities – or effective where a client is looking for straightforward efficiency – fielding guru-like experts when the client wants to follow a prescribed plan. It's the consulting equivalent of mass-marketing: little has been done to establish the client's precise requirements. 'Mass knowledge management' needs to give way to a more CRM-like approach – one-to-one knowledge management – in which consulting firms can tailor the balance of efficiency and effectiveness – of broad and specialised information – to suit the unique circumstances of every project.

---

[1] *Knowledge Unplugged: The McKinsey & Company Global Survey on Knowledge Management*, Jürgen Kluge, Wolfram Stein and Thomas Licht (Palgrave, 2001).

16

Automating Consulting Services: Re-Evaluating Online Consulting

The jury on online consultancy came back a couple of years ago, way ahead of all the evidence being presented, to conclude that the defendant was guilty as charged. Online consulting – by which I mean replacing the face-to-face interaction of conventional consulting with a series of web-based tools enabling clients to conduct the various stages of a consulting project for themselves – was never going to amount to much more than a niche service, provided by a particular segment of the market. Offline consultancies were largely safe: some would develop an online component to their business, but it would never account for more than a fraction of their total revenue.

> *What clients want:*
> Services that match their requirements and offer value for money

> *Value-based consulting:*
> Finding new, more transparent, self-service delivery models that increase, rather than decrease the client-consultant dialogue

> *What consultants want:*
> An on-going personalised relation-ship with important clients

The consulting world breathed a collective sigh of relief: doomsday scenarios of consultants being replaced by expert systems and artificial intelligence networks receded. More than this, the inherent identity of 'consulting' seemed confirmed and strengthened as a result. 'There were quite a few Cassandras out there', said one senior partner I spoke to, 'some of whom were in our own company, all of whom were saying that the Internet heralded the end of consulting as we knew it. And we only had to look to other industries – financial services, for example – to see instances where trusted advisers were apparently being sidelined by the attractions of 24×7 availability and lower costs. But what the last

185

year [2000–01] has shown us, is that personal relationships are difficult to replace. An online system may have advantages in terms of speed of delivery and cost, but clients won't trust it in the same way that they'd trust a consultant they knew well.' 'Online consulting essentially reduces the consulting process to a transaction', agreed another, 'and there's very little of value in that for either the consultant or the client. It's relationships that matter.'

But who really benefits from this relationship? The consulting firm, clearly, because it's more likely to win work when it's stayed in touch with a client, and knows what's happening and how the decision will be made. The firm's cost of sales, too, should be lower, as it has to invest less in research and getting to know the decision-makers. But what does the client gain? Consulting firms would say that clients get a better project – a consultant familiar with the client is in a better position to provide advice on whether a project is appropriately defined and can structure it in a way more likely to deliver the hoped-for results. A cheaper project – consultants who can 'hit the ground running' know how to get things done in the client organisation, they know how to negotiate the rawer political sensitivities. A more successful project, more efficiently executed: who could ask for more? But to what extent is the probability of success actually determined by the nature of the client-consultant relationship? I, like everyone else in the consulting industry, have seen projects sold via a client relationship: some succeed and some fail. Equally, I've seen one sold to a client where there's been minimal interpersonal contact: some succeed and some fail. Indeed, I've seen plenty of occasions when consultants have abused the client relationship to sell in more consultancy than a client needed, and in areas where no assistance was required. Relationships seem at best neutral in their contribution to the delivery of value to clients.

Heretical though it may be to suggest it, perhaps even trust – that keystone of the client-consultant relationship – is an overrated virtue. Surely trust is what clients are often forced to rely on because they have no other means on which to make a decision? The fact that so many surveys reinforce the message that it's the gut-feel of a client when he or she meets a consultant that determines whether the consultant is hired or not, rather than all the quasi-scientific evaluation criteria corporations invent to justify their choices, isn't so much proof of the importance of trusting your instincts, as a testimony to the fact that clients have nothing else to go on. When you buy a packet of breakfast cereal from your local store, you probably do look at the brand, you certainly take into account what you've eaten before and liked and disliked. But you can also look at the ingredients. Do you want a cereal with more nuts and less fruit? With wheatgerm, but not oats? Sesame oil or honey? You don't have to buy on trust.

Now, before you throw this book down in disgust, *of course* I'm not suggesting that buying consultancy is like buying breakfast cereal. But just imagine, for a second, how strange it would be if we did buy breakfast cereal on trust, if every packet was blank except for the faintest outline of its contents. It hardly makes for an ideal breakfast – or an ideal decision-making process. This is equally true in consulting. In fact, I think the situation has been getting better over the last few years: consulting firms are putting far more effort into making the contents of their respective 'packets' as transparent as possible – articulating their experience, their points of view, their approach. But trust is still needed to cover all those intangible elements which a firm cannot illustrate. Trust is what you have when you don't have information and, over the years, we've made a virtue of this necessity.

My point is this. If trust is, in practice, a substitute for something clients want – that the service matches their requirements and transparent value for money – then perhaps we should give the idea of online consulting a reprieve, and examine the ways in which its ability (or otherwise) to deliver on these two fronts may mean that it is capable – at some point – of supplanting the offline client-consultant relationship. Let's therefore look at some of the assumptions that have been made.

Central to these assumptions is the argument that the take-up of online consulting will be confined to small to medium sized enterprises. This way of thinking is based on three more-or-less explicit premises: that smaller companies are more cost-constrained and will choose the low-cost online consulting route in preference to paying substantial fees for conventional consulting; that smaller companies have simpler questions – because they're simpler organisations; and that smaller companies lack the resources – people or information – that enables them to answer these comparatively simple questions by themselves. The converse was also deemed to be true: large corporations had more complex issues which they were prepared to pay a premium to resolve, using their own internal resources to resolve more routine issues. The market for online consulting could therefore be clearly defined. If it was going to expand at all, it would be into more complex questions for those same small-to-medium sized enterprises who, having gained confidence (and value) from posing simple questions, would gradually make more sophisticated demands (Figure 16.1).

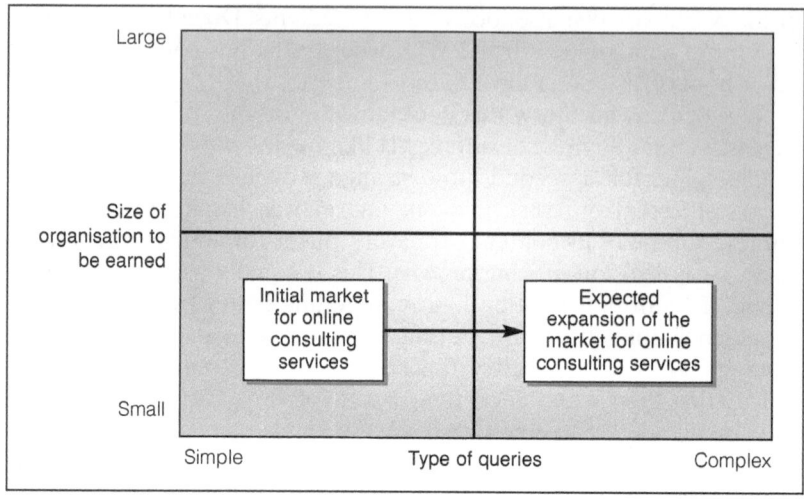

Figure 16.1 *Initial assumptions about the market for online consulting*

The first point we might want to query here is that large organisations are less price sensitive than smaller ones. Clearly, they have larger budgets, and that means that they're capable of spending far more on consultancy in total than a smaller company, but for any given project, they're likely to be just as interested in keeping the price as low as possible. Typically, they can't do that because there's no information available about what they're being charged – there's nothing on the bill from the consulting firm that itemises the overhead of delivering a project in person. Unlike the consumer who can, for instance, compare how much they'll pay for car insurance via a broker and via a direct sales company on the Internet, purchasers of consultancy can't compare a quote from a conventional consultancy with an online one. At the moment, for purchasers of consultancy, the online/offline price differential is not available. If it were – and assuming that the online price would be significantly cheaper – we might find far more clients willing to go down the online route. Would online consulting actually be cheaper? It's difficult to see how it could not be: even where firms charged a premium to cover the development of the intellectual capital to which they are giving clients access, it's hard to see these costs equalling conventional time and materials billing (which, in any case, has a development overhead built in to the fee rate). You could argue that, even if such online/offline comparisons were available, they'd be meaningless because projects delivered by consulting firms would be quite different to ones delivered by clients themselves. It all comes down to the ratio of benefits to the costs: we're accustomed to thinking in 80:20 terms – 80 per cent of the benefits for 20 per cent of the costs. Perhaps, in online consulting, the level of benefits achievable would be lower – 50 per cent for example, but the reduced costs could still be sufficient to tempt many clients – including the large, corporate ones. Clients want demonstrable value for money – and that's something that an online service may be able to deliver more easily than offline consulting.

A second point that's worth querying in relation to the market for online consulting services, is that the consulting needs of very large corporations will be too complex to be answered by an online service which necessarily contains generic information and tools. First, you have to ask whether the level of standardisation of online tools will be significantly higher than those available offline, particularly in the more mature consulting markets where clients are largely seeking to replicate their competitors' initiatives rather than be leading-edge or innovative. As any offline consultant will tell you, an important part of the initial stages of scoping a project is questioning people in a client's organisation, to understand the context and specifics of the issues they've raised, whether these are strategic, operational or technical. The issues of a large corporation are complex because of the number of people involved and their inevitably varying perspectives: in other words, the diagnosis is as complicated as the identification of a solution. Online consulting may not be well-designed when it comes to finding and applying the right solution to the right problem, but it's very well suited to diagnosis – tools that allow people to benchmark themselves against others or which lead people through a decision-making process.

But this is all speculation: what's happening in practice?

Moving to an Online World: Ernst & Young

The US firm of Ernst & Young originally developed a number of different online products and offerings, including: ERNIE, its consulting question and answer system; a US tax Q&A service; accounting and auditing tools; TaxCast, a tax library, and PeopleCast, a human resources tool.

' Different parts of the US firm had been really innovative', recalls
John Odell, who's responsible for rolling out Ernst & Young's
online services in the UK. 'It had seen a need in a particular market
niche and had provided a solution, but this meant that each system
had a completely different look and feel. We got to the stage where
some of our major clients had two, three, four or maybe even five
different online service offerings from E&Y and had to remember
different passwords. A project was launched in 1999, initially called
ERNIE 2000, aimed at pulling these different offerings together. We
wanted to build a brand new platform which would be the one portal
to whatever service offering people want from E&Y.

The platform itself is just a database which points to different tools
and different content. Although generating the content and tools has
been difficult, that's easy compared to the change management side
of things. It's involved significant changes to our business processes,
and so it's been important to make sure that people use and support
the new functionality internally. If we take just one of the tools – for
example, the tax advisor – you can now ask a question online of 22
countries around the world and you're guaranteed your response
back within three working days. If we don't get the answer back to a
client in that time, they get the answer for free. Another distinctive
feature of the new-look Ernst & Young Online is the team rooms for
online collaboration. When we originally launched, these were
extremely limited, not much more than an e-mail directory and basic
discussion thread. We've made a significant investment to improve
this, to facilitate full online collaboration shared in secure spaces,
with shared calendars, tracking of document changes, and all relevant
materials accessible in one place by the authorised users, 24 by 7.
These add up to a real shift in the balance of power, from the
consultant to the client. In the past, if a client was asking a question
over the phone, they could really only ring up from 9am – 5pm;
online, they can do it 24 hours a day. We've also got a freephone help-
desk open 9am – 7pm. Offline, they'd have no idea when they were
going to get an answer back; online, we guarantee the turnaround.
Offline, clients had no idea how much it was going to cost; online, we
guarantee it won't cost more than a specific amount. And, online, you
can ask a question wherever you happen to be in the world – it's not
location-specific. Moreover, in the offline world, you would have to
ring your engagement partner who might, or might not, be an expert
in the question you had, and you wouldn't necessarily know who to
speak to. With the online version, we guarantee to route your

question directly to a top E&Y expert wherever they are in the world.

From a client point of view, there are huge benefits to the online world. However, adoption rates are complicated by the fact that – we believe – we're the first professional services firm to provide online advice on this scale. Many of the people we're targeting – primarily chief financial officers – are perhaps of a generation that didn't grow up using computers so it's not second nature for them to go onto the net to get a solution. That's quite a communications challenge. Some clients understand the concept and take to it straight away; others are just not interested at all, so we continue to deal with them through our traditional channels. In terms of take up, I'd assumed that the smaller clients would be particularly keen on online solutions, but my gut feeling was proved wrong. In fact, I think take-up is high in small companies only because they're perhaps more likely to have younger people working for them – people that are more technologically aware – and have less in the way of resources to answer questions themselves. According to our analysis, some of the multinationals use it just as much as some of the smaller companies. What we can also see is that there's a clear bias towards the more high-tech companies: if they've already got their own Intranet, they're obviously more open to network working. I suspect it all comes down to individuals in companies who are either enthusiastic or not, so it's difficult to generalise on a client by client basis.

Take-up also depends on how well we've managed to sell the idea internally to the engagement team. We've good examples where partners have bought into it internally, so when the client phones up with a question, the partner says, 'I'll tell you what, I'll type this question into our online system and you'll get the answer back through it'. And once a client's seen that the advice available online is as at least as good as that available offline, they start to use it themselves. The nature of the query is also a factor: the simpler the question, the more inclined people are to ask it online. But we also can – and do – handle more complex queries, as people become more comfortable with using the system. Although we've limited the scope of a question to being something that can be answered in less than one day of partner time, so you can get a detailed answer to a highly complicated question in that time. People use the Q&A facilities, but we've found it's the diagnostic tools that are proving most popular.

We're now live in eight countries (although clients in other countries also use the system). We have over 6,000 companies across the world signed up and that's just over 40,000 different users. In the UK, we've got 80 of the FTSE100, compared to fourteen audit clients. Statistics like those are very important internally, showing that we've been able to open doors with clients that hadn't wanted to give us the time of day before, leaving the door open for a future relationship.

The system also has benefits from our point of view: it's more efficient, because we get to see what a client's issues are sooner rather than later.

We probably bill more of our time than we did. Obviously, that's almost impossible to measure, but, in the past, if a client raised a question, time spent answering it often wasn't captured or the client billed, perhaps because the person who answered the question had a specific relationship with the client or perhaps because it didn't really involve much time. Now, there's a charge with every question that comes through, and the person answering it knows they're going to get the credit. It's good, too, from a risk management point of view, there's a very clear record of exactly what's been asked and answered. There's no risk of us misunderstanding a phone question, or of the client misunderstanding what we've said.

Developing such a system, and putting all the supporting infrastructure in place has been a complex and, at times, difficult job. So far as the content was concerned, we went for the easy wins first, incorporating knowledge bases that either already existed or were in the process of being built. This also meant that we could get to market very quickly. But what I'll always regret is that when we first launched internally, it looked as though we only covered certain services. People still sometimes say, 'oh it's the tax thing.' Even when we'd added everything else in, people continued to think, 'I'm not interested in tax so I won't look at it'. Of course, the upside was that we had more positive results to show more quickly: when it came to developing the financial reporting modules, we could say 'look, we've been very successful with this in the last year, wouldn't you like to try a financial reporting version?' I guess there's a trade-off between the brand you want to create and what it's feasible to implement.

To make it work internally, we have to train our staff to see themselves as part of a truly global response network: they've got to prioritise in order to meet deadlines or they're effectively working for free. The advisor network works automatically by sending questions on to 'segment routers', who in turn pass questions on to individual experts within his or her own team. Every answer is then automatically routed to the appropriate reviewer. Everyone's responsible for replying within the deadline set – the whole network's accountable. There's an automatic software escalator process so we can identify any delays, send reminders, and so on. We also ensure that the partners in charge in each of the different countries have fully bought into the process: if they don't then that country doesn't come on board. We haven't yet had to give a free question for missing a deadline, but there'll come a time. Getting them to understand why this is so much better than before, and why it's differentiating us to the market has been one of our challenges. We've been going to different parts of the business, telling them what's been successful. It's primarily been a proactive approach; sometimes we do come across online initiatives that are just being started, so we can say, 'that's great but we've already got the infrastructure and the help desk in place and the admin. Let's not reinvent the wheel.' We also held a huge number of face-to-face meetings with individuals, especially those who didn't feel comfortable with our approach.

Use of the system has yet to be mandated, that's something I'd quite like to do, but it would be too unpopular and we're therefore trying the carrot rather than the stick. There are various incentive programmes and every month all the people that have asked an online question get put into a hat and one person receives a prize – a trip in a hot air balloon or a day at a health spa. We try and make it fun and to get people talking: because with so many initiatives across the firm, all of which are important in their own way, it can be difficult to get air time.

Where do I see E&Y Online going in the future? To date the technology hasn't been advanced enough, robust enough, or secure enough for us to be confident about being able to post and exchange documents online. We just couldn't take the risk that unauthorised people could gain access to highly confidential documents. We're piloting a way of doing this, and have almost reached the point where we can offer this to everyone, and that's going to mean a significant change to the way of working, for clients who want to work like this, and certainly for our people internally. Everything will be much faster. Already on the corporate finance side, online collaboration is used to quite a large degree, saving the whole paper-chase of faxing and couriering documents around between ourselves, the merchant bank and other third parties. To be able to have everything online and be able to see who's made what changes, represents a significant improvement in efficiency. We'll also be looking to make more 'interpretive' knowledge available, rather than just factual queries and responses, some of the thought leadership the firm generates, for instance. It all comes down to calibration: whatever the issue is, we can input expert knowledge from different parts of the world within a very short space of time. Now that we've got the processes in place to do this, and changed our culture to deliver it, we can really add value to the way we work with our clients. **,**

Automating Consulting Services: Towards a Value-Based Approach

Taken together, clients – even large corporate clients – may well feel that value for money is more demonstrable and the match between needs and solutions more exact with online consulting (Figure 16.2).

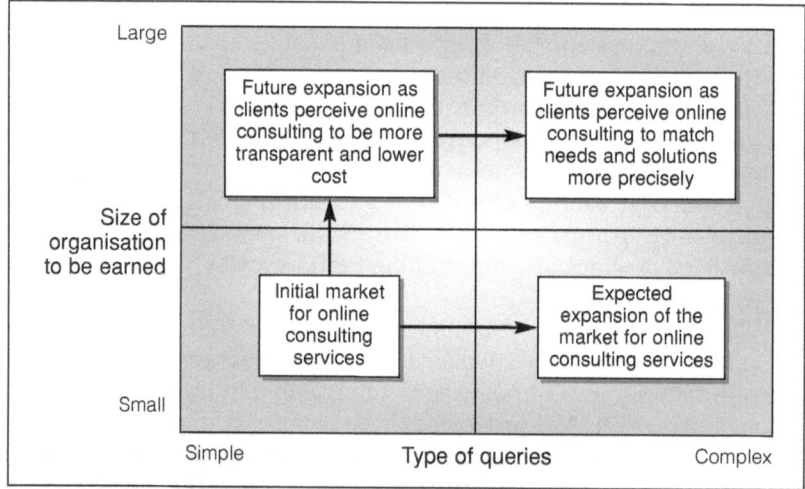

Figure 16.2 *Refining the assumptions about the market for online consulting*

In considering what intellectual capital should be made available online, many firms started at the solutions side, putting up reports, surveys and methodologies from work that had been completed. Clients may be interested in this, but there's an overhead for them in filtering the plethora of available methodologies down to the one they need.

Investing more in diagnostic tools might represent a more valued form of online intellectual capital – and the experience at Ernst &

Young reinforces this. Furthermore, by moving the use of these tools from the consultant (who may have a vested interest in selling a particular solution) to the client, clients may be able to demonstrate to their own satisfaction that the solution a consulting firm is offering is the one they need. In other words, online self-service diagnosis may offer an advantage over the conventional offline relationship, where a client may well be suspicious of a consultant's intent. Crucially, it may also offer a means by which the traditional, face-to-face client-consultant dialogue is strengthened, not cannibalised.

17

Automating Delivery: The Route to Business Transformation

But the technocratisation of consulting isn't limited to those perhaps increasing number of occasions where the content of consulting can be viably offered online. More and more aspects of the consulting process are becoming technology-dependent.

Take marketing consultancy as an example. The type of market research required in a multi-channel environment would appear to be a huge opportunity for consulting firms: in effect, it's the logical extension of the market analysis that has been the mainstay of many a consulting engagement in the past. Not surprisingly, therefore, many con-

> **What clients want:** Consulting firms to be making appropriate use of technology to streamline their operations and cut out inefficiencies

> **Value-based consulting:** Using technology to control a multi-dimensional business

> **What consultants want:** To improve efficiency, but not at the cost of heterogeneity and individuality

sulting firms have developed proprietary research, intended to act both as a door-opener into potential clients and as a means of positioning themselves on the revenue-generation – rather than the cost-reduction – side of the industry. Similarly, just as the technology firms have been quick to exploit the perceived strategic importance of technology to move into more general business consultancy, so are market research companies starting to offer consultancy support.

But the real opportunity for consulting firms to contribute in this area may be much more limited in practice. While proprietary research may, indeed, have an important role to play in establishing a firm's overall intellectual credentials, there's considerable scepticism on the part of clients about the value of conventional market research techniques in a mulit-channel world. 'Traditional market research – surveys, panels, focus groups – can't tell us what we really needed to know, which is how customers would actually behave faced with the combination of proposition and technology we were offering', said one. Another client recalls going live with a web-site but then winding it down after a month because they weren't getting either the traffic or quality of response they wanted. We then carried out an in-depth statistical analysis of our visitors – who they were, where they came from, how they behaved when they entered the site – and then spent six months completely redeveloping the concept. If we hadn't done this, I'm in no doubt that we'd have failed. Monitoring traffic on a website is the online equivalent of following shoppers around to see what they buy and don't buy and what they do on each return visit.'

Companies faced with the problem of trying to second-guess customers' behaviour have two routes available to them. First, they can disintermediate the research company and go straight to the customer, involving the latter during the proposition development phase. 'If you're really going to "own" your customers across multiple channels', said one interviewee, 'then you have to take a holistic approach. The market is moving continuously, so you can't afford to turn away from it. You also need a fundamentally different attitude at board level, because you can never expect to stand still.' Second, they can launch a prototype site and use one of the sophisticated site-usage monitoring tools now available to analyse how customers are behaving in practice. Around 65–70 per cent of online consumers – whether they're individuals or corporate purchasers – abandon their shopping carts before they get to the checkout: to understand why this happens, you have to start a dialogue with users and capture information you gain as part of the process. Using intelligent software will become increasingly important as companies launch more and more variations around their core product set. Such opportunities require a fundamental change in the way companies sell their products and services. As one person I spoke to put it, 'there has to be a shift towards seeing products from the customer perspective, so that you can capture their preferences and understand the trade-offs that you are making in choosing to meet – or not meet – specific needs. You have to use technology to do this: it's not something you could understand through conventional market research. But the software has to behave like an expert salesperson,

by enabling people, for example, to build their own virtual car.'
'Companies are going to have to invest more heavily in technology-
based marketing management because it's the only way they'll be
able to demonstrate their cost-effectiveness', said another.

But, at the same time, there's concern that some of the fundamental
lessons learnt in traditional marketing have been ignored in the switch
to a multi-channel environment. 'We find many companies', said one
consultant, 'that just haven't applied the kind of monitoring and
analysis they carry out for their conventional campaigns to online
campaigns. It's as though they thought that the old rules don't apply
here. But they do: more so than ever.' The situation is complicated by
the fact that, because of the complexity and fragmentation of the
technology market in this area, it's been the IT department, not the
marketing department that's been making the key decisions and, in
large part, running projects. This should change as more top-level
people are appointed who can bring different sides of the organisation
– principally the marketing and IT functions – together. 'The most
successful clients we encounter', said another consultant, 'are those
that have managed to integrate all elements of their marketing mix
and been able to reapply the learning from one area to another very
quickly. These are the kinds of companies that are asking "Do we
know where we make our money?" Deciding whether or not to
embrace a new channel isn't just a question about technology: it also
goes to the heart of whether you as an organisation have the capacity
to cope with the resulting complexity and fluidity. And it helps, when
you're engaged in this kind of discussion, to have an outside view. As
a client, in this kind of marketplace, you can't expect a generalist
consultant to know everything. Equally, as a consultant, you have to
come clean and say where your own expertise ends: you have to
involve specialist companies.'

But the hybrid offline/online world of future marketing poses
consultants who specialise in that area with some specific challenges.
As in the clients they serve, there's a danger that the role of the
traditional marketer will be subsumed into that of the technologist.
What value is there in strategic marketing plans which aren't based
around in-depth understanding of the impact of technology on
consumer behaviour? How can you evaluate – let along optimise –
advertising expenditure if you can't interpret the detailed output
from electronic channels? In the course of a handful of years,
marketing consultancy has evolved from a service not dissimilar to
strategy, to something more akin to IT consultancy.

Moreover, this external pressure is being mirrored by internal pressures. The consulting industry has been slow to think of itself in terms of a position on a supply chain. The reasons for this were largely cultural: the operating unit of even the largest consultancies has primarily been one person – the consultant. Rather than seeing consultants as part of a chain, they've been positioned more as central points in a hub. The support functions of a firm are designed to do what their name suggests – support the consultant, who was, in effect, an end in his or her own right. The idea of a supply chain challenges this conceptual model: instead, it places consultants in a large processor, where they not only consume inputs (the support services of the collective firm) but produce outputs (projects that have to be evaluated, clients who have to be surveyed, bills that have to be paid – and so on).

Similarly, project delivery was perceived to be the work carried out by the consultant, and did not necessarily include the start or the end of projects – sales and marketing activities, and post-engagement reviews, customer satisfaction surveys, billing and working capital management. Up to the early 1990s, consulting firms were, in fact, classic examples of the 'silo'-based organisations which they were trying to re-engineer with their clients. Consulting was one stream of activity; collecting the money from that work was another stream entirely. Consultants were responsible for the successful delivery of a project: whether or not a client paid for it was a matter for some anonymous person in finance. By the early 1990s, part of this attitude was starting to change. While their clients were replacing their vertically-structured organisations with flatter ones in which processes were horizontally integrated across different business functions, consulting firms took a different route. Essentially, this involved putting more functions in the hands of the consultant, rather than trying to integrate the activities of the consultant with other parts of the business. This shift was dressed up in the guise of greater accountability – consultants had to realise that clients didn't just have to be satisfied with the support they received, they had to pay for it. But it was also an admission that consultants tended to be lone operators, accustomed to working in independent teams, rather than repeatedly coming back to the office for assistance.

What makes this picture more complicated was that, at the same time that firms were investing in supporting individual consultants, the collective entity of the firm was growing in importance. Several factors drove this change: the greater investment being made by firms in marketing and brand promotion, in knowledge management and systems, and in global infrastructures; a growing interest in exploring alternative ownership models; the number of strategic alliances being formed. All of these gave the corporation hiding within the conventional consulting firm much greater visibility – something that was always going to come into conflict with the strategy of making the consultant the centre of everything. That this attitude is now changing is evidenced by the fact that it used to be support staff who were cut when times were hard; by contrast, many of the cutbacks announced by consulting firms in 2001 were of consultants.

So what we had, by the end of the 1990s, was a situation in which the strategic direction of the firm (towards centralisation) was almost diametrically opposed to its operational reality (decentralisation, down to the level of the individual consultant). And this was in an environment in which clients themselves had been fighting to take out transaction costs (first through ERP systems, more recently through web-based initiatives such as e-procurement) and which increasingly wanted consultancies to swallow a little of their own medicine.

The challenges beyond traditional ERP

So how can a consulting firm reconcile the two sides of its increasingly schizophrenic personality? In a sense, this division has become a self-fulfilling prophecy. Despite the mounting pressures – internally and externally – to automate parts of the consulting process, we've ended up in a situation in which consultants can only ever be seen as individualists, and technology as uniform. Solving this problem clearly involves moving away from such a stark dichotomy: moving away from the idea that one system equals one view on the world.

It's a view shared by Per Tejs Knudsen, the Chief Executive of Maconomy, a Copenhagen-based company developing integrated software to control service delivery.

❝ Globalisation is polarising the market into two types of firm – the very small and the very large. You can either specialise in a very specific field: such firms will be very like advertising agencies in that they'll be very dependent on individual people – the 'A'-team of creative directors, the small number of people who really understand the business, surrounded by a whole host of support staff. At the other end of the spectrum, you've the full-service firm. Full-service firms try to 'own' client accounts: they may offer some services that aren't profitable just in order to ensure that, whatever a client needs, they're in a position to provide. They want to be able to do everything. For these firms, globalisation is critical: if you want global client accounts, you need to have a global presence.

Each of these types of firm needs to be managed differently. If you look at the small specialised firms, infrastructure's not very important to them because, although they will charge for services, their business is essentially person-dependent. Charging consists of saying, 'okay, I did this for this client, and that should cost $100,000'. But the big firms are what I'd call a production machine, and I think, in their thinking, they'll get closer and closer to the manufacturing sector. For people in manufacturing, the job is to produce items – their competitive focus is therefore on how they utilise their machines. Apply that thinking to the services industry, and you start to think of your people as your machine – and that, in turn, affects the way in which you think about automating your operations. Manufacturing companies bought ERP systems because they wanted to support their business processes, because they wanted to be able to share data, and because they wanted to know how their business was doing at any moment. Once you move to automating service industries, the implications are more complex: technology ceases to be something that simply supports your business, but has to go hand-in-hand with your strategy and organisational design. To optimise your business, you have to be in control of your business. And I believe that that's what global management consulting firms will have to do in the future.

We've seen the beginnings of it, already, with knowledge management. Many firms have implemented Lotus Notes and similar systems in order to share data across practice areas. But these initiatives have been largely confined to exchanging documents, reports, information on clients, and so on. Firms have been much more reluctant to extend automation to include business processes. Part of that's cultural, but a lot of it stems from the recognition that conventional ERP systems were designed for moving data around single organisations. With consulting firms, we're talking about virtual organisations, with people spread, not only in geographical terms: some are on customer sites, some are at home, some are in offices. To support the virtual organisation you need a different kind of infrastructure – and that's where the Internet becomes really important. What you want are common business processes and a single database which enables you to share common data throughout your whole organisation. You have to have a common infrastructure going across the world which allows you to look at your business from different perspectives – according to different dimensions, in a sense.

You'll have a geographic dimension, of course, but you may also have partners who own 'communities' that cut across countries. You'll want to know how, say, France is performing, but you'll also want to look at, say, your Oracle services worldwide. And you might have the partner in the telecommunications practice working for a bank customer, so – again – you'd have to understand performance from both perspectives. The key is to get all of this online within a single system; then, as soon as you've got the timesheets in, you can see how well you're leveraging your related, how you're performing in practice against budget, and so on.

We're currently working for global consultancy where the system has to be able to report data on a country-by-country basis, for legal reasons, but where there's a considerable amount of transferring people between countries in reality. They don't just want a system that can collate data in only one dimension: they need to look at how different business units and industry teams are performing. By the time we've finished, there'll be several thousand management consultants on one system.

The implications for this client's business are substantial. One of their internal objectives is to remove boundaries because their current business is, in reality, a multitude of smaller businesses in which each partner owns his or her little kingdom. With this technology, they've become one, virtual organisation. They're still reporting into a country for regulatory reasons, but this is of no importance for them: the real business owners are the people responsible for industry consulting practices, who may be based anywhere. Another big change is in terms of internal pricing. There may be an official price that you're aiming to achieve, the sales price (which is what you actually achieve, the transfer price which allows you charge back the use of resources internally, and the cost price (how much you actually pay for a resource). These prices are hugely influential in driving behaviour. Take transfer prices as an example. If you set it too low, then the business from which a resource is being charged out will be less willing to make its people available to other business units: financially, they'd do better to focus on winning external work, where the margin will be better. But set the transfer price too high, and the other business units will be reluctant to use internal resources: they'll prefer to go outside the firm and protect their margins by using cheaper labour from other companies. So the transfer price becomes a political tool: suddenly you can manage behaviour, simply using your system to do it. You can create a P&L down to the level of the individual employee.

Moreover, by putting everyone onto the same scale, you highlight the differences between countries and industries. Typically, even the most global of global consultancies find that they've been operating multiple price lists, in effect. One industry may be doing well, while another is in the doldrums – and that has a knock-on effect in terms of the kind of rates you might be aiming to achieve and those that you're achieving in practice.

The other big impact is on an organisation's business processes. Once a firm has got its financial model right, it can manage its organisation much more from a financial point of view. It's usually the case that different country operations – even in global firms – have different business processes. But having a single financial model means that you can prevent individual partners having their own policies; you can now take a regional or global approach and say 'this is how we operate'. This is crucial if the largest consulting firms are to be able to share common services across countries and achieve genuine economies of scale. By standardising the whole infrastructure, the collective 'entity' of a firm has a chance to evolve into a kind of headquarter service.

The theory is one thing: the question is how to make it work in reality.

When we talk to clients, their main concern is whether this can be done. There are two sides to making projects of this type work in practice: the organisational side and the IT side. What we're really talking about here are business transformation projects: we're helping clients move from being what I call multi-local to being global. That's a business transformation – and one of the most important things is that clients themselves recognise this. The decision to implement a system like this has to be taken at the highest levels within the consulting organisation, and it has to be a decision that everyone – not just a single champion – buys into. At the same time, it needs a sponsor who combines clout (remember the saying that no one ever put up a statue to a committee) with a relentless attention to detail. Issues like the ones I've mentioned – reporting structures, pricing details – are all hugely important, but you do find people who try and delegate this kind of work to junior staff, and that – at least in our experience – just doesn't work. You also have to recognise that you may be coming from a partner environment where everything is democratic to a more centrally-controlled one. People have to accept that this means that you are a professional management team and that the partners' power is diminished. Partners will still be shareholders, but they will have lost some of their clout. From an organisational point of view, there has to be a centralised approach: it has to be seen as a business transformation project.

From an IT point of view, there needs to be a recognition that an ERP-type strategy simply won't work in this environment. We're talking to one firm at the moment that's got almost 15,000 change requests logged for its planned ERP system: it's clear that no-one really believes these can be done. You really have to have web-based software to be able to handle the complexity, heterogeneity, and openness of the consulting organisation.

Sometimes we find companies that don't want to contemplate what they see as being a tremendous disruption in terms of implementation. Typically, this is driven by the age of the management team. If you've earned your money, have done well and are getting ready to retire, you think why buy an extra headache? You don't really earn any points with your colleagues for spearheading these kinds of initiatives. To be honest, I think there has to be some sort of catalyst – a burning platform, if you will. Mergers and acquisitions are perhaps the most obvious examples. When you merge two similar sized companies, there'll be two sets of everything, so integration becomes an absolute priority. Interestingly, PR companies have adopted systems like this much more readily, and we think that's partly because that sector's gone through such a tremendous period of consolidation and restructuring. The consulting industry is just beginning a similar process – and we believe that will provide the impetus that automation has hitherto lacked. And we expect that the firms to emerge from that **?** process will be managed very differently indeed.

Automation: Towards a Value-Based Approach

Conventional ways of automating consulting firms have largely been – and continue to be – monolithic. Such systems are never, in fact, likely to yield the kind of improvements in efficiency desired by clients. Instead, because they're seen as compromising the historical autonomy of the consultant, resistance to their implementation has been considerable at all levels.

Moving to a more centralised structure and more uniform business processes has to be matched by developing both a front-end and a reporting structure capable of being used by different people in different ways: capable, in essence, of mass customisation.

Part 4
Conclusions

18

The Value-Based Consultancy

There's a lot of unfinished work around the consulting industry.

Alan Buckle, COO, Europe, KPMG Consulting

So what, in summary, will the value-based consultancy look like? How could firms align their own objectives with those of their clients more effectively?

First, in terms of enabling clients to choose the right firm, the value-based consultancy:

	Will balance clients' need to...	With its own imperatives to ...	By ...
Intellectual capital	Have a clear understanding of the areas in which a consulting firm excels so that it can match that competence aganst its own requirements	Retain a market positioning that can be adapted as markets appear and disappear	Being precise about the both the client and consultant's intellectual capital in order to generate long-term client trust – that actually increases revenues, not subtracts from them – rather than falling into the trap of selling short-term management fads
Market intelligence	Be approached only by consultants who aren't selling them 'junk' services, but know exactly what they want, when they want it	Maintain a steady stream of work, enabling it to survive in good times and bad	Supplementing conventional account management processes to make better use of 'pull' technology and establish more of an on-going dialogue between the consultant and client
Brand consistency	Have a guarantee (of sorts) that what they see is what they get	Control its exposure to the potentially variable performance of individual consultants	Creating and living a brand capable of reinventing itself in response to changing client needs

Second, when it comes to providing clients with what they need in terms of consulting services, value-based consultancy:

	Will balance clients' need to...	With its own imperatives to ...	By ...
Specialisation	Have access to world-class expertise in specific areas	Retain a flexible organisational structure, capable of responding to changes in clients' needs	Reducing the intervention of the corporate firm, to enable individual consultants to develop and switch specialisations
Skills integration	Have specialist consultants working seamlessly together	Have high staff utilisation	Being able to provide an environment in which integration takes place, without direct intervention
Innovation	Feel confident that they're in the forefront of management thinking (not to look stupid); to be seen as innovators – but in a comparatively low-risk way	Be able to maintain the level of innovation required by clients, but without compromising their profit margins	Creating organisations in which innovation is part of the structure, not an adjunct to it
Infomediation	Have access to the unique information gathered/analysed by the consulting firm	Continue to control the intellectual value chain	Focusing on developing unique customer insights with practical application
Structured methodology	Have guarantee (of sorts) of certainty in terms of what a particular consulting service will deliver	Achieve economies of scale by being able to re-use the same methodology for multiple clients	Reducing the dependency of both clients and consultants on overly-structured methods
Networking	Have access to the reach of a consultant's network, without ceding power	Play a leading role in creating new businesses and markets, and to be seen as movers and shakers, not just observers	Clients and consultants working together to create new business opportunities
Change management	Do what they cannot do themseves	Undertake work that it is posssibe to do successfully	Finding new models for facilitating change as an organisational model

Finally, when it comes to performing its work in the right way, the value-based consultancy:

	Will balance clients' need to...	With its own imperatives to ...	By ...
Organisational design	Work with consulting firms that are able to reconfigure their resources to meet their clients' requirements	Have manageable organisations	Having an organisational structure in which flexibility is balanced with the need to exert control
Value-chain integration	Have access to world-class skills	Be able to access new markets and manage sudden spikes in demand by building relationships with software vendors and specialist firms	Clarifying the role that the consulting firm plays, and by integrating the skills of different partners around specific propositions
From theory to practice	Have a seamless transition from business strategy to technology implementation	Be able to provide a seamless service which neither cannibalises its fee rates in strategic work, not becomes prohibitively expensive when it comes to implementation	Finding new ways to resource projects while maintaining an integrated service from the client's perspective
Knowledge management	Be ablel to accecss the collective intellectual capital of conssulting firms, as well as that of individual consultants	To reduce the firm's reliance on individual consultants	Adopting a one-to-one approach, identifying a client's needs in terms of formal and informal knowledge
Automating consulting	Have services that match their requirements and offer value for money	Have an ongoing, personalised relationship with important clients	Finding new more transparent, self-service delivery models that increase, rather than decrease the client-consultant dialogue
Automating delivery	Work with consulting firms that make appropriate use of technology to streamline their operations and cut out inefficiencies	Improve efficiency, but not at the cost of heterogeneity and individuality	Using technology to control a multi-dimensional environment

Cumulatively, these characteristics point to several trends, likely to become significant in the consulting industry in the future.

If there's one trait that almost all the companies profiled in this book share, it's their willingness to do to themselves what they do to clients. They represent a very significant shift from the kind of consulting firm that offered theory not practice, that chose to prescribe solutions they would not apply to themselves. Part of this reluctance 'to do as you would do to' stems from the reasonable view that every company is different: solutions that worked for a client wouldn't automatically work in a consulting environment. Medicine for the goose is not necessarily medicine for the gander. But, as boundaries between industries have crumbled, creating new opportunities and threats, as more clients have launched their own consulting firms, this justification has sounded increasingly hollow. 'It may well be', say clients, 'that this specific solution works for me and not for you, but part of your job as a consultant is take ideas from sector to sector, from organisation to organisation. Why does that process stop in your own atrium? What is so very different about you?' The firms profiled here don't think there's any difference: indeed, they use the fact that they have to get clients to work in a certain way to drive and sustain change in their own businesses. This is something that goes beyond empathy, beyond those hackneyed phrases about 'working in partnership': where clients and consultants face the same challenges and experience the same pain, the traditional distinction between client and consultant – between them and us – begins to dissolve. And that, in turn, paves the way for different, more open ways of working.

Another factor that comes through strongly here is the changing role which the corporate entity – 'the firm' – will have to play. Having experienced something of a renaissance in recent years – thanks to the pressures to create global infrastructures, the need to invest in technology and new product development – the firm as a collective entity needs to reinvent its role. When it comes to organisations, people and processes, the direct intervention of a firm may create more problems than it solves. But a recurring theme here is the need for more effective but less direct interventions on the part of the corporate firm – allowing consultants to respond almost instantly to a changing client, for example.

But, of course, it's not just a question of *allowing* people to do this. What's much more important – and, again, this is a theme that's re-iterated in many of the firms discussed here – is to provide tools that *enable*. In this, technology will be critical: so many of the ways of working discussed in the preceding chapters require more collaborative working, greater exchange of information, more transparency and better communications. While these may be highly sought-after attributes from a client's point of view, they could be prohibitively expensive from the consultant's. Consultancy has grown up on the back of being a people business, channelled almost exclusively through face-to-face communication. Clearly, that aspect is not going to go away in the foreseeable future – indeed, faced with a more volatile, complex business environment – it will be more important than ever. But, that being said, it's not been a model of unalloyed success from the client's point of view: it's high cost; difficult to maintain; its consistency is difficult to regulate; it can become too standardised. There has to be a tremendous competitive advantage for the firm who combines the existing model with technology that allows consultants to do more for less. What more could a client really want?

Appendix:
Information on Contributors

Part 2 – The Right Firm

Chapter 2

Intellectual capital: articulating your portfolio

Fjord was formed in 2001 with investment from Swedish company Spray AB. The company combines digital consultancy services with a number of proprietary software products to build better relationships. At the heart of the company's approach is a comprehensive understanding of the relationships and interconnections which exist in organisations – and how they are affected by the introduction of new technologies, products or services. Fjord uses insights into how relationships are affected by digital media to help businesses improve service, attract new customers and retain motivated employees, and to build digital products that release real value from such relationships.

Based in London and Stockholm, Fjord's major clients are in the sports, telecoms, financial services and media sectors.

Mark Curtis is currently a managing partner at Fjord. Before that he was EVP Skills and Practice for Razorfish globally. He sold his company, CHBi, to Razorfish in 1998 having established it as a pioneering digital services outfit in 1993 with Mike Beeston. At CHBi/Razorfish, Curtis worked with clients like Yell, RAC, Allied Domecq and Sonera. He also led a successful change programme across all 2000 employees worldwide. Prior to this he set up an innovative media promotions company in 1989 and also has a background in sponsorship and PR.

Chapter 3

Market intelligence: avoiding junk mail consultancy

Bob Borsch is the partner responsible for PricewaterhouseCoopers' MCS Global Client Relationship Management Program and is a member of the PwC Global MCS Executive Board. He formerly served in various leadership roles in Coopers & Lybrand Consulting, including Managing Partner of the National Accounts Program, Managing Partner of the Consumer Products and Retail Consulting Group, Managing Partner of the Midwest Consulting Region, and Chairman of the Coopers & Lybrand Global Manufacturing Industry Program. Before joining Coopers & Lybrand, Borsch held numerous regional and local consulting leadership roles in Ernst & Young from 1977 to 1989, primarily in manufacturing, retailing and aerospace, as well as office and regional practice management roles. He has a degree from Western Michigan University in Accounting and Finance.

Scient is an e-business consultancy which uses its extensive e-business experience to reduce cost and create revenue opportunities. Since its founding in 1998 Scient's only business has been e-business, from strategy development through to implementation. The differentiated approach which clients come to Scient for is a blend of strategy, customer experience (branding, fulfilment and usability) and technology.

Randall McComas is the General Manager of Business Development for Scient. He joined the company as the General Manager of the Global Telecommunications and Utilities Business Unit, where he advised telecommunication and utilities clients in establishing e-business solutions to create market leadership, breakthrough positions, and economic results through e-business strategy and execution. Prior to joining Scient, McCormas was Vice President of Telecommunications for IBM Global Telecom and Media industries. While at IBM, he held various positions in the telecommunications and media industry, including Vice President of Marketing and Strategy, where he managed IBM's worldwide telecom business unit including wireline and wireless carriers and ISPs. Prior to joining IBM, McCormas was Vice President of Marketing and Engineering at Kline Engineering. He also worked as a registered structural engineer in South Carolina and performed nuclear submarine duty for the United States Navy. Randall has a BS in civil/structural engineering from The Citadel. He completed IBM's Harvard Executive Education Program in 1991.

Chapter 4

Brand consistency: delivering experience as well as services
Since being established in 1913, Andersen has enjoyed uninterrupted year on year growth by developing innovative solutions which help

people and organisations create and realise value. Andersen offers Business Consulting, Global Corporate Finance, Tax and Legal and Audit and Business Advisory services, with 390 offices in 84 countries.

Ashley Unwin is the global head of the People and Change Practice at Andersen, with responsibility also for the practices in Europe and the Middle East. As a partner in the Andersen organisation's Business Consulting Practice, he specialises in providing advice to clients on the subject of organisational change. His experience covers the management of large-scale change initiatives within both the public and private sector, ranging from leading edge software organisations, global media organisations and global retail organisations. A large amount of Unwin's experience has been acquired supporting strategy implementation, technology-related change, business transformation, integration and acquisition. Over the last six years, Unwin has led and been involved in a number of initiatives, including the development of Business Consulting's Change Enablement Framework, which supports all of Andersen's change projects. He has a BA in Business Studies and an MSc in Organisational Behaviour from Sheffield University, UK.

Chapter 5

Specialisation: letting market forces prevail

Bain & Company is one of the world's leading global business consulting firms, serving clients across six continents. It was founded in 1973 on the principle that consultants must measure their success in terms of their clients' financial results. Bain's clients have out-performed the stock market 3 to 1. With 2,800 employees, head-quarters in Boston and 27 offices in all major cities throughout the world, Bain has worked with over 2,000 major multinational and other corporations from every economic sector, in every region of the world.

Steven Tallman is a Vice President at Bain & Company and splits his time between the firm's Brussels and San Francisco offices. In addition to having led client teams in a broad cross-section of industries, he has global operations responsibility for the firm's Knowledge, Training and Technology Services. Since joining Bain in 1986, Tallman has worked in Bain's San Francisco, Moscow, Warsaw, Tokyo, Boston, London and Brussels offices across a variety of industries. These include telecommunications, financial services, consumer products, electric utilities, and entertainment. Projects have included re-engineering, business unit strategy, privatisation, and market entry strategies.

From 1997–2000, Tallman served as Vice President of Training, responsible for designing and implementing a worldwide training and development strategy. He had oversight responsibilities for 20 global training programmes and developed Bain's cutting edge 'Virtual University' to facilitate customised training for each employee.

In 2001, Tallman became Vice President of Global Services. He currently has global operations responsibility for Knowledge Management, the Global Experience Center, the Information Services function, MIS and Technology, Worldwide Training Programs, the Bain Virtual University, the Bain.com website, Web Services, the Client Communications and Graphics function and coordination of the firm's Practice and Capability areas. Tallman is a graduate of the University of California, Davis, where he received a Bachelor of Arts, *summa cum laude*, in Economics and Political Science.

Chapter 6

Skills integration: an unworkable model?

Marakon Associates is an international consulting firm that works with top management to deliver corporate strategies and capabilities that maximize shareholder value. The firm is built on the premise that the long-term interests of all stakeholders are best served when management can create the highest value for investors over time. It was established in 1978 by three former corporate finance executives at Wells Fargo Bank and a distinguished academic. Marakon's original headquarters was in San Francisco; it was shifted to Stamford, Connecticut in 1987.

Dominic Dodd is a Managing Partner at Marakon, based in the London Office. His consulting experience is in the financial services, consumer goods, retail and telecommunications sectors, working in long-term relationships to provide a broad range of strategic and organisational advice to chief executives and senior management. He is a member of Marakon's Global Managing Partner Team and has responsibility for the firm's Corporate Communications and Branding Strategy.

Founded in 1993, Inforte is a demand chain management consultancy, helping organisations integrate strategies, processes and systems across the value chain in order to create an integrated, customer focused and demand-driven approach to planning and executing operations. Inforte provides the strategic thinking, technology enablement, systems integration and user experience that makes this possible and effective for clients around the world. Inforte is headquartered in Chicago.

Philip Bligh founded Inforte with the vision of providing value to global companies by maximizing the convergence of business strategy and emerging technology at every point along the value chain. He serves as its chief executive officer and chairman of its Board of Directors and is responsible for guiding the company and providing strategic vision for its growth. Inforte is one of the USA's leading and fastest growing strategic technology consultancies. Bligh is an author and frequent speaker on the subject of strategy technology topics and e-Business issues. Recently he contributed research to the first of three articles on the future of the Internet. The article, written by Harvard Business School Professor Michael Porter, was published in the March 2001 edition of the Harvard Business Review. The next two articles are scheduled to be released later this year. Bligh graduated with a Bachelor's degree in Chemical Engineering from University College of London.

Chapter 7

Innovation: getting more for less

Differentis is an e-business integrator which helps organisations realise value from their information technology investments by forging linkages to create a 'joined-up' business. To do this, Differentis acts as: a counsellor, helping senior management sort out the legacy of past IT initiatives and develop a clear direction for IT; an architect, understanding how the business can be enhanced by IT linkages; a constructor of technology; and a manager of reliable resources who implement the Joined Up Business.

Founded in 2000, Differentis' expertise has been gained from working with Europe's leading organisations. They employ a unique operating model that is designed to continually improve their ability to rapidly deliver solutions that meet business requirements. The organisation's commercial proposition promotes an efficient high quality working relationship that ensures Differentis' work meets its clients' objectives and the firm is paid for the work they produce, not the time they consume.

Ron Mackintosh, Differentis' Chief Executive Officer, has spent over thirty years in the IT industry including the last ten as one of the top executives in Computer Sciences Corporation. From 1992–2000 Mackintosh was President and Chief Executive Officer of CSC's European business. During that time CSC's European business grew from $180 m to $2.5 bn in revenues and employed over 16,000 people. In leading this he was actively involved in creating CSC's consulting business, making seven major acquisitions across Europe and in securing Europe's first $1.5 bn commercial outsourcing contract. Before joining CSC Mackintosh spent ten years with Nolan Norton and Co. and Index Group, two firms that played a lead role in shaping the management and strategic approach to using IT for competitive advantage.

Bruce J Rogow is an independent information technology executive counsellor, commentator and mentor, currently working as a Differentis 'fellow'. Throughout his career as a consultant, executive educator and counsellor, he has been at the forefront of creating the body of knowledge related to directing computer resources. Rogow began his career with IBM, where he spent five years in marketing and five years in Advanced Technical Training, as one of the early developers of what became systems management. His primary responsibility is to assist Differentis and its clients with the development of leading management frameworks, practices and methodologies. He helps clients capitalise on the information technology opportunities and challenges facing them in the coming decade. Rogow also spent a number of years as Executive Vice President of Consulting, Research and Executive Services for the Gartner Group and he will continue to support Gartner Executive Program Research and European clients in an advisory role.

Nick Shelness, another Differentis 'fellow', is widely recognised as a thought leader in the IT industry, and an expert in the fields of computer based collaboration and messaging. He served as the Chief Technology Officer (CTO) of Lotus Development – an IBM company from 1998–2001. Prior to serving as Lotus' CTO, he held other senior technical positions at Lotus and Soft-Switch Inc. The latter, which he joined in 1984, was acquired by Lotus in 1994, prior to Lotus' acquisition by IBM in 1995. He was named a Lotus Fellow in 1996, elected to the IBM Academy of Technology in 1998, and named an IBM Fellow in 1999. From 1970–1980, Shelness was a Research Assistant, and subsequently a Lecturer (US equivalent, Assistant and Associate Professor) in the Department of Computer Science of the University of Edinburgh, where he led early research in multi-computer operating systems.

Paul Seaton is one of Differentis' venture directors and, as such, specialises in the creation and shaping of ventures in the initial stages of their evolution. Prior to joining Differentis, he was an Associate Director of CSC, where he specialised in devising innovative solutions enabled by emerging technologies. Seaton has extensive retail and travel experience and has undertaken many business process change and emerging technology integration projects.

Chapter 8

Data, information and knowledge: re-engineering the intellectual value chain

As one of the world's leading corporate strategy firms, Mercer Management Consulting helps leading enterprises achieve sustained shareholder value growth through the development and implementation of innovative business designs. Mercer's proprietary business design techniques, combined with its specialised industry knowledge and global reach, enable companies to anticipate changes in customer priorities and the competitive environment, and then design their businesses to seize opportunities created by those changes.

The firm's capabilities and leading-edge intellectual capital are enhanced by its industry expertise and geographic range. For 30 years, Mercer has worked with leaders of major companies in chemicals and pharmaceuticals, communications, computing, consumer goods, financial services, health care, media and entertainment, manufacturing, oil and gas, retail, transportation and utilities. The firm also offers special services in the areas of Internet strategy and private equity investing.

Mercer Management Consulting is part of Mercer Consulting Group, one of the world's largest consulting organisations. Together, the firms of Mercer Consulting Group have 14,000 employees in more than 30 countries throughout the world.

Rick Wise leads Mercer Management Consulting's North American strategy practice. He has particular expertise in helping large companies identify, prioritise and exploit new growth opportunities through innovative business designs. Specific project work has included helping clients develop and launch new businesses, structuring and evaluating strategic alliances and acquisitions, developing innovative go-to-market and channel strategies and leading the redesign of outdated business strategies. Wise has worked with leading companies in the automotive, manufacturing, financial services, retail, software and materials industries, among others. Wise played a key role in developing Mercer's Value-Driven Business Design process for systematically identifying and exploiting opportunities for companies to create shareholder value growth. He was a significant contributor to Mercer's recent books on growth strategy, *Value Migration, The Profit Zone* and *Profit Patterns*. He holds a BA from the University of Pennsylvania and an MBA from the Wharton School.

Richard Balaban is a Vice President in the strategy practice of Mercer Management Consulting and has been based in its London office since it opened in 1982. He has worked on business and corporate strategy studies in a wide variety of industries, including building materials, oil products, specialty chemicals, telecommunications, consumer products, paper and forestry, financial cards and payment systems, and airlines. His work has ranged from comprehensive strategy reviews and implementation projects to more focused profitability improvement, manufacturing cost improvement and acquisition and investment planning studies. Balaban has extensive experience with the design and application of implementation programmes and tools matched to particular business development and investment initiatives. He graduated in history from Kenyon College and holds a master's from Stanford University in European history. He received an MBA at the Wharton School of the University of Pennsylvania.

John-Paul Pape is a Vice President within Mercer's Strategic Capabilities Group, where he specialises in all aspects of valuation, shareholder value analysis and value-based management. Pape has run engagements in industries such as banking and finance, pharmaceuticals, chemicals, oil and gas, retail, leisure, paper and heavy manufacturing, ranging from the full implementation of shareholder value principles within a firm to the valuation of a firm for the purpose of a transaction such as an acquisition, disposal, joint venture or Initial Public Offering (IPO). Prior to joining Mercer, Pape led the value-based management practice of a boutique consulting firm. Pape holds a Bachelor's degree in business Studies with a specialisation in finance and shareholder value analysis from Middlesex University in the UK.

Chapter 9

Structured methodology: the consulting prisoner's dilemma
Darrell Rigby, a Director at Bain & Company, heads the annual management Tools & Techniques research at the Boston-based global

business consulting firm. He specializes in corporate strategy and global retailing practices. With 20 years of management consulting experience, Rigby has led assignments in a wide variety of industrial and consumer industries including high technology, health care, retailing, and financial services. For the past eight years, he has directed the global survey of senior executives to gather facts about the use and performance of management tools. Each year, he has published the *Management Tools Guide* containing the top 25 most popular and pertinent tools. This guide has been widely cited in the business pages of many US and international publications, including *Forbes*, *The Financial Times*, and *The Economist*.

Prior to joining Bain, Rigby earned an MBA from Harvard Business School with high distinction (Baker Scholar). He is a graduate of Brigham Young University where he received a Bachelor of Science in Business Management *summa cum laude*.

Chapter 10

Networking: adding value in a joined up world

American Management Systems (AMS) is an international business and information technology consulting firm that helps clients create value by increasing revenues and market share and by decreasing costs. The firm combines expertise in business analytics, business process design and information technology with a deep understanding of the industries it serves, including financial services, new media and communications, energy, healthcare and US federal, state and local government.

AMS has specialist expertise in the areas of customer relationship management; credit, market and operational risk management; e-business; billing and operational support systems; enterprise and B2B integration. The firm's clients are industry-leading organisations and AMS derives over 85 per cent of its business each year from clients with whom the firm has worked in previous years. Founded in 1970, AMS is headquartered in Fairfax, Virginia, with 8,750 employees and 51 offices worldwide. AMS reported 2000 revenues of $1.28 bn.

David Yates is General Manager for AMS in Northern Europe with overall responsibility for all of AMS client engagements in that area. Prior to taking on this position, he was head of the AMS' Corporate Banking Practice based in New York, where he was engagement manager for a variety of high value client engagements in the area of corporate banking business process re-engineering, business case development and consulting for in- and outsourcing of processing businesses. Yates is also a member of the Management Board and the AMS Engagement Manager for Proponix. Yates has eighteen years of experience consulting to and carrying out major systems implementations for blue chip financial institutions worldwide, in which capacity he has acted as engagement manager on projects ranging in size from one year product implementations with a value of $3 m to multi-year systems integration projects in the context of corporate mergers and organisational renewal with a value in excess of $40 m.

Bill Graham is the President and Chief Executive Officer for Proponix. In this role, he provides overall leadership in shaping the global business management and operations for this new outsourcing venture, which provides trade services processing to international trade banks. Prior to joining the Company, Graham had a successful career at Citibank.

Most recently, he served as Asia-Pacific Regional Head of Financial Institutions, based in Singapore, where he was responsible for an operation covering 13 countries and 165 employees. Graham serves on the Dean's Advisory Board of the Schulich School of Business, as well as the advisory boards of both the International MBA program and the e-business program at the Schulich School of Business, York University, Toronto. He has an MBA with distinction in International Finance and Business from the Schulich School of Business, York University and a B.Sc. in Mechanical Engineering from Queen's University, Kingston. He is also a Professional Engineer.

Part 3 – The Right Way

Chapter 12

Organisational design: delivering solutions, not services
IBM Global Services is the world's largest information technology services provider, with nearly 150,000 professionals serving customers in 160 countries and annual revenue of more than $33 bn (2000). IBM Global Services integrates IBM's broad range of capabilities – services, hardware, software and research – to help companies of all sizes realize the full value of information technology.

Mike Howarth is Director of Marketing worldwide for IBM's Business Innovation Services - the consulting and systems integration division of IBM Global Services. He has been with IBM for 24 years in a range of technical, management, sales, and marketing leadership roles in the UK, in IBM's European Headquarters in Paris and in Worldwide headquarters in New York. He has been a pioneer in the interpretation and execution of classic marketing techniques into the world of professional services in several different units. During 2002 he is returning to work in the UK in a European-wide IBM role. Howarth holds a degree in Physics from St Catherine'is College Oxford and a Diploma in Management from Henley Management College.

The Virtual Development Group is an organisational and management consultancy working at strategic and operational levels specialising in strategic thinking and implementation using virtual working methodology; it was founded in 1995 and is based in the UK. It concentrates on providing consulting solutions to complex client needs; coaching and mentoring; facilitation of client meetings and workshops; organisational design and change management programmes; and bespoke development work. Clients include The Boots Group, British Telecom, Environment Agency and Unilever. Part of the company's long-term vision is to create a group of businesses based around virtual working.

Martin Clemmey is a Director of The Virtual Development Group and a consultant specialising in strategy and finance. He is a Chartered Accountant, having trained with PricewaterhouseCoopers where he gained experience of working with major multinational clients both in the UK and abroad. On leaving the profession he worked with a US based Fortune 500 company and then joined Halma plc, a Top 250 company, as Financial Controller. Martin has an MBA in Strategic Management from Henley Management College and his most recent corporate line role was as a Director of Halma Process Safety Division. This was a part line, part consultancy role involving strategic, marketing and financial management of companies in the UK, Europe, USA and Australia.

Chapter 13

Value chain integration: Building constellations
IconMedialab is a leading global e-business consulting and integration firm. Utilising user-driven solutions, developed with leading technology and solid, time-tested business processes delivered anywhere in the world, we help clients build stronger, more profitable relationships with customers, business partners, employees, suppliers and shareholders. The company was founded in Stockholm in 1996 and now has more than 30 offices in 15 countries including Australia,

Belgium, Denmark, Finland, France, Germany, Italy, Netherlands, Norway, Portugal, Spain, Sweden, Switzerland, UK, USA.

Erik Sandersen is a citizen of Norway. He is currently VP Global Partnerships for IconMedialab and part of the international management team. Former experience includes Managing Director of Circle Innovation, a Norwegian IT consulting firm, and Manager and Case Leader for the Boston Consulting Group in London and Oslo. Sandersen holds a Master's degree in Computer Science from the Norwegian Institute of Technology and an MBA from Stanford University.

Cap Gemini Ernst & Young is one of the largest management and IT consulting firms in the world. The company offers management and IT consulting services, systems integration, and technology development, design and outsourcing capabilities on a global scale to help traditional businesses and 'dot companies' continue to implement growth strategies and leverage technology in the new economy. The organisation employs about 60,000 people worldwide.

Paul Nannetti is a Vice President at Cap Gemini Ernst & Young, and the European Leader for CRM and DareStep. Nannetti joined Cap Gemini in October 1994 and during his first two years held the position of UK Financial Controller. In 1997 he became Business Process Management Director and developed what was then a fledgling business from £4 m to £40 m p.a. in two years, seeing an increase in staff from 100 to 1000. In 1999, Nannetti took board responsibility for developing Cap Gemini UK's approach to e-business. His division, the E-Business Unit, was responsible for driving innovation and urgency in Cap Gemini's response to e-business, and for taking leading edge solutions to market. In September 2000, Paul was appointed Managing Director of DareStep in Europe. DareStep is a division of Cap Gemini Ernst and Young providing user-centred solutions for the Internet and other electronic channels, integrating strategy, design and technology competencies. This division now has 700 people globally, the majority of whom are based in Europe.

Nannetti was appointed to his present position, European leader of CRM and DareStep, in April 2001. In this role he is responsible for the development of the CRM business across Europe, including the definition of offers and propositions and practice building, together with sales and delivery. Prior to joining Cap Gemini Ernst & Young, Nannetti worked for IBM Europe, and started his career as a Chartered Accountant with Ernst and Young.

Chapter 14

Technology: managing the transition from theory to practice

Silverline Technologies is a leading international e-business and integration services firm, employing over 2,400 software professionals worldwide, as of 30 June 2001.

Silverline was incorporated in 1992 and has traded publicly on India's Bombay Stock Exchange since 1993. In June 2000, Silverline Technologies became the first Indian Technology company to be listed on the New York Stock Exchange. The company provides a comprehensive set of e-business consulting and IT services including strategic consulting, creative design, technology integration and implementation, as well as management and maintenance of Internet and legacy applications, focusing primarily on Global 2000 clients. Silverline delivers its services through a global network of software development centres. At the heart of the network are core offshore centres in Chennai, Hyderabad and Mumbai, in India. These centres support regional development facilities located close to clients throughout North America, Europe and Asia Pacific.

Chris Baker is Silverline's Senior Vice President and Managing Director of Silverline Europe, with responsibility for the development, growth and management of Silverline's business in Europe. He has over 20 years' experience in business, having held senior management positions in manufacturing, IT solutions and consulting firms. With SeraNova, prior to that company's acquisition by Silverline, Baker was responsible for the start-up of SeraNova Europe, building SeraNova's European team and winning the company entry into the European market, with a number of strategic client wins. Before joining SeraNova, he held board level and senior positions in manufacturing, most recently with Automotive Products plc, in Information Technology for CSC Computer Sciences Corporation, and in management consulting for Ingersoll Engineers and Coopers & Lybrand. Baker holds a BSc (Hons) in Production Engineering and Production Management from the University of Nottingham, and an MSc in Machine Tool Technology from the University of Birmingham in the UK.

Dan Hayter is Silverline Europe's Executive Vice President of Sales and Marketing for Europe. Hayter has spent his entire career in the IT industry and currently heads up Silverline's European Sales operation. In this role, he has led the build-up of the company's European business in application maintenance outsourcing and e-business application development, as well as customer relationship management (CRM) by developing a strategic alliance with Europe's leading CRM software vendor, AIT. Prior to joining Silverline, Hayter worked at KPMG and Unisys. At KPMG, Hayter was instrumental in setting up their new systems integration business; at Unisys he developed and ran the company's European eCRM programme.

Chapter 15

Knowledge management

Intellectual Capital Sweden AB was founded in March 1997. With thoughts and theories of intellectual capital as a starting point, a model for valuation of knowledgebased companies has been constructed. From this model, a tool has been developed – IC Rating™ – which measures intellectual capital and makes it comparable between companies and between units within a company, and which has been validated through empirical analyses of a large number of companies within IT, finance, communication, media and management-consulting.

The company started as a small unit for development but has grown extensively since then. Today it consists of graduates from Schools of Economics, Behavioural Science and Systems Engineering.

Leif Edvinsson is one of the world's leading expert on Intellectual Capital. As the world's first corporate director of Intellectual Capital at Skandia of Stockholm, Sweden, Edvinsson has been a key contributor to the theory of intellectual capital and oversaw the creation of the world's first corporate Intellectual Capital Annual Report. In the light of his work, Edvinsson has been a special advisor on service trade to the Swedish Ministry of Foreign Affairs. He is also special adviser to the Swedish Cabinet on the effects of the new digital economy, a special advisor to the United Nations International Trade Center and co-founder of the Swedish Coalition of Service Industries. Edvinsson holds an MBA from the University of California, Berkeley. He is the author for numerous articles on the service industry and on Intellectual Capital.

Chapter 16

Automating consulting services: re-evaluating online consultancy

Ernst & Young UK is an integral part of the Ernst & Young world-wide organisation, a global business committed to being the trusted business advisor who contributes most to the success of people and clients by creating value and confidence. Ernst & Young's broad inventory of services and solutions is delivered on an integrated basis to clients by 80,000 people in more than 130 countries. Ernst & Young Online is the connection between Ernst & Young's clients and its people and knowledge worldwide. Ernst & Young Online was launched in the summer of 2000; 80 per cent of FTSE 100 companies are now registered users of Ernst & Young Online. Services offered include: news alerts and analysis, reference library, collaborative services through a 'communicate with Ernst & Young' facility, Ernst & Young perspectives on business issues, and online tools, such as the 'accountants' bible' UK & International GAAP*plus*.

John Odell is Project Director for Ernst & Young Online. After graduating from the University of Oxford with a Master's degree in Engineering Science in 1995 (where he specialised in computing options), Odell joined Ernst & Young to train as an auditor. After four years in Audit, Odell was chosen to work for the UK Executive on the firm's e-business strategy and he co-wrote the feasibility study for Ernst & Young Online. He now leads a team of twenty people, focusing on using Ernst & Young Online as a differentiator, and using it to work more effectively with clients, promoting efficiencies, enhancing communications, building new relationships, and adding new depth to existing ones.

Chapter 17

Automating delivery: the route to business transformation?
Maconomy provides integrated business software for the leading companies in the world. Based on an innovative technology, Maconomy delivers net-based solutions that optimise service delivery, relationship management and e-commerce as well as the accounting and financial functions that support the global organization and integrate the entire service value chain. Customers include Avnet, Millward Brown, Avis/Cendant, IBM, Deloitte & Touche, KPMG, Rambøll, Transco, Xerox Industry Solutions & Services (XISS), and Philips Design. Maconomy has offices across Europe and the US.

Per Tejs Knudsen, one of Maconomy's founders, has served as the Group's Chief Executive Officer and a member of the Board of Directors since the Company's formation in 1988. In 1983 Knudsen founded PPU Software A/S, now a financial holding company and one of Maconomy's major shareholders. Knudsen has been the guiding leader of Maconomy: his vision of developing solutions for service-industry automation has been validated by the strong acceptance that Maconomy solutions enjoy across services-industry segments. As the CEO of Maconomy, Knudsen has led the company through several product cycles by identifying unmet market needs and leading Maconomy's development of new solutions to meet those needs. Knudsen received a MSc degree from the Technical University of Denmark in 1982 and a B Com degree from the Copenhagen Business School in 1988.

Index